The CRC Press
International Series on Computational Intelligence

Series Editor
L.C. Jain, Ph.D.

L.C. Jain, R.P. Johnson, Y. Takefuji, and L.A. Zadeh
Knowledge-Based Intelligent Techniques in Industry

L.C. Jain and C.W. de Silva
Intelligent Adaptive Control: Industrial Applications

L.C. Jain and N.M. Martin
Fusion of Neural Networks, Fuzzy Systems, and Genetic Algorithms: Industrial Applications

H.N. Teodorescu, A. Kandel, and L.C. Jain
Fuzzy and Neuro-Fuzzy Systems in Medicine

C.L. Karr and L.M. Freeman
Industrial Applications of Genetic Algorithms

L.C. Jain and Beatrice Lazzerini
Knowledge-Based Intelligent Techniques in Character Recognition

L.C. Jain and V. Vemuri
Industrial Applications of Neural Networks

INDUSTRIAL APPLICATIONS of GENETIC ALGORITHMS

Edited by

Charles L. Karr
L. Michael Freeman

CRC Press

Boca Raton London New York Washington, D.C.

Library of Congress Cataloging-in-Publication Data

Industrial applications of genetic algorithms / edited by Charles L.
    Karr and L. Michael Freeman.
        p. cm. -- (International series on computational
    intelligence)
        Includes bibliographical references and index.
        ISBN 0-8493-9801-0 (alk. paper)
        1. Evolutionary programming (Computer science) 2. Genetic
    algorithms. 3. Artificial intelligence. I. Karr, C. L. (Charles
    L.). II. Freeman, L. M. (L. Michael). III. Series.
    QA76.618.I55 1998
    005.1—dc21   98-30681
                                                                            CIP

No claim to original U.S. Government works
International Standard Book Number 0-8493-9801-0
Library of Congress Card Number 98-30681
Printed in the United States of America 1 2 3 4 5 6 7 8 9 0
Printed on acid-free paper

CONTENTS

CHAPTER 1

GENETIC ALGORITHMS IN THE ENGINEER'S TOOLBOX

Charles L. Karr and L. Michael Freeman
Department of Aerospace Engineering and Mechanics
University of Alabama
Box 870280
Tuscaloosa, AL 35487-0280
e-mail: ckarr@coe.eng.ua.edu and mfreeman@coe.eng.ua.edu

ABSTRACT

Genetic algorithms (GAs) are computer-based search techniques patterned after the genetic mechanisms of biological organisms which have allowed such organisms to adapt and flourish in changing, highly competitive environments for millions of years. These robust genetic algorithms have been successfully applied to problems in a variety of fields of study, and their popularity continues to increase due to their effectiveness, their applicability, and their ease of use. This chapter provides an overview of genetic algorithms including a brief history that indicates GAs have made the leap from their origins in the laboratory and the halls of academia to the practicing engineer's toolbox. The GA applications presented in subsequent chapters are meant to serve as a testament to this idea. Each chapter describes a project completed by a graduate student as an assignment for a semester course offered at The University of Alabama. The current chapter describes The University of Alabama course, and lays a foundation for the rest of the book.

INTRODUCTION

Almost from the beginning of the computer age, researchers have been interested in developing computers that demonstrate the adaptive capabilities of natural organisms. Early computer scientists such as Alan Turing and John von Neumann were interested in developing what is commonly referred to today as artificial intelligence: machines (computers) capable of understanding, adapting to, and controlling the environments in which they exist and function. Many of these early pioneers had backgrounds in biology, psychology, and medicine, and thus looked to natural systems as guides in the pursuit of their goals. Over the years, these efforts have produced techniques and systems of varying effectiveness, complexity, and popularity such as expert systems, neural networks, fuzzy logic, adaptive agents, and evolutionary systems.

Today, computer methods inspired by biological evolution are grouped under the umbrella of a field called *evolutionary computation*. The three main elements of evolutionary computation are

1. evolution strategies [1],
2. evolutionary programming [2], and
3. GAs [3].

Each of these three techniques mimic the processes observed in natural evolution, and provide efficient search engines by evolving populations of candidate solutions to a given problem. These techniques have proven to be effective in solving difficult problems from a wide range of disciplines including economics [4], engineering [5], medicine [6], and chemistry [7], to name a few. GAs are generally thought to be the most prominent technique in the field of evolutionary computation.

GAs utilize an iterative approach to solving search problems. They are population-based search techniques that rely on the information contained in a broad group of candidate solutions to solve the problem at hand. This population-based search distinguishes GAs from traditional point-by-point search routines. The strength of GAs lies in their ability to implicitly identify the inviting properties associated with potential solutions, and to produce subsequent populations of candidate solutions which contain new combinations of these fertile characteristics as derived from candidate solutions in preceding populations.

GAs code the requisite information for a solution to a given problem in strings called *chromosomes*. Each chromosome can be decoded, according to a user defined mapping function, yielding specific values for each of the important parameters being sought. Coding schemes can vary dramatically from one GA application to the next, and in fact, have themselves undergone somewhat of an evolution. In early GA applications, most coding schemes were designed to produce chromosomes that were bit-strings consisting of concatenated binary sub-strings designed to represent the individual parameters necessary to form a solution to a particular problem. More recently, researchers have moved toward coding schemes which represent the various solution parameters with floating point chromosomes, resulting in chromosomes that bear a striking resemblance to arrays common to most computer languages. For many GA novices, it is this coding of the problem into chromosomes that represents the most difficult conceptual hurdle. However, once a potential GA user has been exposed to some GA applications, it is not long until said user is able to develop imaginative and effective coding schemes for a particular problem. The coding scheme plays a key role in the ultimate success or failure of a GA application.

The potential solution represented by each chromosome in the population of candidate solutions is evaluated according to a fitness function which is synonymous with the objective function of traditional optimization: a function that quantifies the quality of a potential solution. This fitness function is used in the implicit identification of high-quality values of the individual solution parameters, and the goal of the GA is to either maximize or minimize the fitness function of the strings in a generation depending on the specific problem. The fitness function ultimately determines which chromosomes are selected to propagate their parameter values through subsequent generations. Like the coding scheme, the fitness function plays a key role in the GA's success (or failure) in any given problem. It has been said that a GA will find what it is told to look for as represented by the fitness function defined. Thus, as the old adage goes, "be careful what you ask for, you just may get it."

New populations of candidate solutions are generated by implementing operators inspired by natural genetic variation. The three most popular operators used in almost all GAs, are:

1. *selection,*
2. *recombination* (often termed *crossover*), and
3. *mutation.*

Selection is a process through which high quality candidate solutions are chosen to form a basis for subsequent generations of solutions. Selection operators are generally driven by probabilistic decisions that ensure the best solutions are given the greatest consideration. *Recombination* is an operation by which the attributes of two quality solutions are combined to form a new, often better solution. This operator plays a key role in determining how the quality attributes of the candidate solutions are combined. *Mutation* is an operation that provides a random element to the search. It allows for various attributes of the candidate solutions to be occasionally altered. Together, these three operators produce an efficient search mechanism that generally converges rapidly to near-optimal solutions.

Given a coding scheme, a fitness function, and specific genetic operators, it is rather straightforward to develop a GA that mimics natural evolution to effectively drive toward near-optimal solutions. Despite the fact that the particulars of any GA application may vary, the basic approach is the same. The fundamental algorithm is summarized as follows:

1. Generate a random initial population of candidate solutions in the form of chromosomes.

2. Evaluate each chromosome in the population according to a pre-defined fitness function.

3. Employ a selection operator to create new chromosomes. The selection operator biases the new generation of chromosomes toward higher quality solutions. As the chromosomes

mate, genetic operators such as recombination and mutation are applied to form new candidate solutions.

4. Delete members of the existing population to make room for the new candidates.

5. Evaluate the new chromosomes and insert them into the population.

6. If a satisfactory solution has been achieved (or if some other stopping criteria has been met), stop; otherwise, go to step 3.

This approach to solving search problems may seem unusual at first consideration. However, GAs demonstrate at least three characteristics which allow them to perform efficiently in difficult search spaces. First, GAs consider a population of solutions simultaneously. Thus, they tend to maintain a global perspective due to their parallel consideration of solutions, whereas traditional search routines tend to perform local searches due to the point-by-point manner in which they traverse a search space. Second, GAs do not rely on derivative information. Thus, they tend not to get fooled in multi-modal search problems that perplex derivative-based techniques so popular in the field of optimization. Third, GAs do not depend on continuity of the search space. Thus, they have been used to solve problems that have stymied more traditional search routines which can break down when applied in discontinuous search spaces.

Although GAs have been used to solve a wide variety of problems, most GA applications fall into one of three categories: (1) optimization, (2) machine learning in the form of learning classifier systems, and (3) genetic programming. The optimization applications are characterized by evaluation functions that are indicative of some desired result. Classic examples of optimization problems are those in which some profit measure must be maximized or some error term must be minimized. However, GAs have expanded their horizons well beyond traditional optimization. A second use of GAs is in learning classifier systems which employ a GA for developing effective if-then rules for solving a particular machine learning problem. GAs are used in these systems as discovery and adaptive engines that drive a rule base toward an effective solution. A third use of GAs is genetic programming [8], an approach to problem solving in which a GA is used to discover entire computer programs. Candidate programs are represented as tree-like symbolic expressions (S-expressions), and a GA breeds existing programs to develop new, more effective programs. As always, the search for more efficient computer programs is driven by a fitness function.

GAs and evolutionary algorithms of all types can be quite fascinating and are growing in popularity. However, they are by no means some "fly-by-night" approach to problem solving; GAs are well grounded in theory, and empirical studies demonstrate that, in some problems, they consistently outperform more traditional search routines. Although GAs are a very young technology, their brief history is rich with success stories and peppered with researchers from diverse backgrounds. The next section provides a flavor of this rich history and mentions some of the researchers who have elevated GAs to their current status in the computational community.

A BRIEF HISTORY

As with most technologies, the specifics of GA history are open to debate. This is due in part to the difficulties associated with distinguishing the lines of demarcation between GAs and other evolutionary algorithms. Therefore, this section is intentionally not restricted strictly to a discussion of GAs, rather it attempts to credit early researchers who developed techniques that served as a foundation upon which today's GAs have been built.

Perhaps the earliest instance of an evolution-based technique related to modern GAs was suggested in the 1960's by Rechenberg [1]. This technique was called *evolution strategies* (*Evolutionsstrategie* in the original German), and was originally used primarily in an attempt to design airfoils. It provided a mechanism by which real-valued parameters were altered to optimize an airfoil shape based on an objective function. This work was continued by Schwefel [9, 10] who served as a focal point of a reasonably active community of researchers. However, the "populations" in evolution strategies consisted of but two candidate solutions: a parent and an offspring which was a mutated version of the parent. For the most part, GAs are thought to have developed independently from evolution strategies. It has not been until recently that the two research communities have begun to interact significantly.

Evolutionary programming is a technique developed by Fogel, Owens, and Walsh [2] which is actually quite similar to modern GAs. In evolutionary programming, candidate solutions to a problem are represented as finite-state machines. New solutions are evolved by mutating the candidate solutions (the evolution is accomplished by randomly altering the state-transition diagrams). The production of new candidate solutions is driven by a fitness function. Unlike GAs, evolutionary programming uses only a mutation operator to provide variation in the population of candidate solutions.

Several other researchers in the 1950s and 1960s developed evolution-inspired computer algorithms. These efforts have not received the attention or follow-up that evolutionary programming and GAs have, but are worthy of mention nonetheless because they addressed the main application focal point of GAs, that being optimization and machine learning. Box [11], Martin and Cockerham [12], and Bledsoe [13] all developed algorithms that were based on the search capabilities apparent in natural systems. Additionally, there were a number of natural biologists including Baricelli [14] and Fraser [15, 16] who used computers early on to simulate evolution in very controlled experiments. However, these efforts did not address the more general optimization and search goals of GAs.

John Holland is generally credited with the invention of GAs. Holland, his students, and his colleagues at the University of Michigan developed a detailed approach to modeling natural evolution in the form of computer algorithms. Holland's 1975 monograph entitled *Adaptation in Natural and Artificial Systems* describes the basic approach to the population-based search characteristics of today's GAs. His book presented most of the genetic operators used in GAs including selection, recombination, mutation, and inversion. Interestingly enough, Holland's original motivation was not optimization, rather it was to formally study the mechanisms of adaptation found in nature and to incorporate those mechanisms into computer-simulated systems. One of the real appeals of Holland's work is that it set forth a theory attempting to explain why and how GAs worked. Holland's *schema theorem* is still the most popular theory used to explain the success of GAs. In the years following the publication of Holland's book, his students struck the greatest blows for GA acceptance through the publication of empirical studies in which GAs were used to solve optimization and machine learning problems [17-20].

The publication of John Holland's book on adaptation in artificial systems represented a high water mark in the history of GAs. His book was the first document in which anyone had set forth the idea of using a population of candidate solutions in conjunction with a suite of genetic operators. A second major milestone in the history of GAs came with the publication of David Goldberg's 1989 book entitled *Genetic Algorithms in Search, Optimization, and Machine Learning* [21]. This landmark text presented the GA as a straightforward approach to solving search problems of various kinds. It presented the GA as a problem-solving tool, and cited numerous instances in which researchers had applied GAs in various problem domains. The book also provided a clear and concise explanation of the theory of GAs. Pascal source code for most of the routines presented in Goldberg's book allowed other researchers to experiment with GAs with a minimum of difficulty.

The publication of Goldberg's text accelerated the application of GAs. The extent to which GAs progressed in the application domain is evidenced by the publication in 1991 of *The Handbook of Genetic Algorithms* [5] by Dave Davis. The first several chapters in this book served as a tutorial on GAs, providing the reader with the basic tools of GA implementation, but neglected the theoretical details. The remainder of the book consisted of chapters outlining GA applications. These subsequent chapters were written by individual researchers who had successfully applied a GA in their particular fields of interest. This book served as a testimonial to the applicability and the effectiveness of GAs in a number of problem domains and represented the state-of-the-art in GA applications. However, it also noted a problem at that time: most of the chapter authors were from academia or research firms, or were otherwise GA specialists.

The interest in GAs is definitely growing. In recent years there have been a number of texts published on genetic algorithms [22-24]. Additionally, there has been a very successful biannual international conference on genetic algorithms, and GAs have received growing attention in the popular press. The publicity resulting from this growing body of literature has contributed to a dramatic increase in GA applications. The number of publications related to GAs is not only growing, it has virtually exploded over the last decade. Figure 1.1 represents a reasonable approximation to the number of GA papers published annually [25]. The bibliography of GA applications compiled by Alander [25] gives an indication of the diverse fields in which GAs are being applied. Table 1.1 supplies a breakdown of the most popular application areas addressed by the papers appearing in Alander's bibliography. This classification by subject is interesting because a large number of the papers are related to engineering applications. It is apparent that more and more engineers are using GAs to solve their problems.

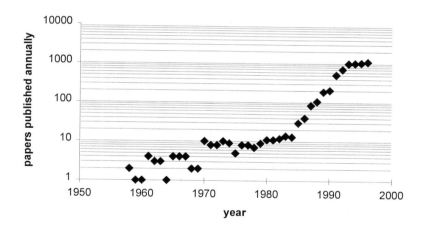

Figure 1.1 – The number of GA papers published annually has continued to increase exponentially.

As the number of published GA papers has grown at such a rapid pace, so too has the number of applications in which GAs have been used to solve industrial-scale problems. Many of these problems are being solved by other than just researchers at universities and government labs. Indications are that GAs are being widely used by practicing engineers and scientists. This expansion in GA users is due to the technique's growing reputation in both the computational and engineering communities and to the increasing availability of information on GAs. What was once an abstract concept that proposed the use of evolution-based operations to provide computers with adaptive capabilities similar to those in nature, has developed into a well-defined computational technique applicable to the solution of a wide range of search and optimization problems.

Table 1.1 – The most popular subjects as addressed by papers in the bibliography compiled by Alander [25].

subject	number of papers
classifier systems	16
review	14
machine learning	13
control	11
classifiers	10
engineering	8
optimization	7
fuzzy controllers	7
learning	6
immune systems	6
adaptation	6
fuzzy control	5
GA analysis	5
computer aided design	5
others	247

In the early days of GA development, the application of a GA to a particular problem generally required consultation with a "GA expert," someone well versed in the intricate details of GAs. Although it is a workable arrangement, this collaborative effort tends to lengthen the process of producing quality results because the GA expert has to be brought up to speed on the application domain; that is, to acquire an expertise that the person or group originally requiring a solution already has. Today, a more desirable situation exists. GAs can be successfully applied to a particular problem by the person or group interested in solving the problem; a GA expert is no longer required to produce quality results. The GA has made its way into the practicing engineer's toolbox. This is evidenced by the fact that first-year graduate students can complete GA application projects with a minimum of introduction to GAs.

A CLASSROOM AT THE UNIVERSITY OF ALABAMA

The College of Engineering at The University of Alabama offers a one-semester course on genetic algorithms which is taught by an instructor from the Department of Aerospace Engineering and Mechanics. This semester course consists of forty-five 50-minute lectures, and is generally offered to graduate students (although seniors are allowed to take the course with the instructor's permission). The prerequisite is one computer course. Although the course is offered in the College of Engineering, the students who have taken the course come from very diverse backgrounds. For instance, the students who wrote the remaining chapters in this book have backgrounds in a variety of fields including engineering, computer science, and finance.

The goal of the course is to provide students with a fundamental understanding of GAs sufficient to use them to solve problems. In addition, students are exposed to state-of-the-art topics in GAs including real-valued parameter representations, genetic programming [8], learning classifier systems, GAs in fuzzy systems, GAs in neural networks, and advanced operators. Goldberg's book [21] is used as the textbook, but the book is heavily supplemented with handout material. The course is divided into the following four main topic areas: (1) basic mechanics of a GA including an introduction to computer code in three computer languages (3 weeks), (2) the theory of GAs (3 weeks), (3) advanced operators (4 weeks), and (4) GAs in machine learning (5 weeks). The students are evaluated based on two 50-minute examinations, one 2 ½ hour comprehensive final examination, a variety of homework exercises, and a semester computer project. The computer projects are proposed by the students, and they are as diverse as the students' backgrounds and interests. It is these computer projects that comprise the remainder of this book.

The computer projects are assigned in the second week of the semester, after the students have been introduced to the fundamental capabilities of GAs. Thus, they are encouraged to suggest search, optimization, and machine learning problems without having been exposed to the vast potential of GAs. Many of the students have selected applications or problems related to their thesis topics; some have subsequently chosen to extend their applications to become their theses. There are few restrictions placed on the computer projects. Students are allowed to use GA computer code made available to them by the instructor, to acquire GA code from outside sources including the Internet, or to write their own (few actually select this option). As in most industrial applications, the majority of the students' time is spent developing effective fitness functions and coding schemes. Many of the coding schemes the students suggest cause them to develop specialty genetic operators. Most seem to be confident in their ability to apply a GA to interesting and complex problems when they finish the course.

It is important to note a special characteristic of the GA course offered at the University of Alabama: the course is offered to numerous off-campus students. The University of Alabama has been a leader in distance learning, developing numerous technology-enhanced classrooms that are used to teach courses which are video-taped and made available to students through a program called Quality University Extended Site Telecourses (QUEST). This program allows students, often working professionals interested in obtaining graduate degrees, exposure to the same lectures offered to students on site. In addition, the GA course is made available to students at the National Technological University (NTU) who are nearly always working professionals. The inclusion of off-campus students is important to note because many of the students take the course with the goal of using a GA to solve a specific search or optimization problem; many of the projects represent real-world problems of interest to industry.

Despite the particulars of the course offered at the University of Alabama, it is not dissimilar to GA courses at numerous universities.

ABOUT THIS BOOK

The editors' motivation in producing this book is to demonstrate the applicability of GAs by users who have had a minimum exposure to the technique. Most of the students have a limited knowledge of GAs, if any, when they begin the course. Since they formulate the solutions to their applications after the first six weeks of the course, they provide a benchmark for what a user can achieve using a GA after a moderate introduction.

The chapters that comprise this book are extended versions of the students' semester projects completed during the 1996 Fall semester. There are three exceptions: (1) Chapter 12 written by Barry Weck represents a project completed as a GA class project in a prior semester, (2) the work described in Chapter 16 by Brown Cribbs was not done in conjunction with the GA course, but was done as a graduate project while Mr. Cribbs was a student at the University of Alabama, and (3) the work described in Chapter 8 by Angie Reichert was completed as a special problems course while Ms. Reichert was an undergraduate student. Although it is true that many of the chapters include extensions to the semester projects, the majority of the work presented was completed during the semester. Much of the additional effort put forth by the students was spent in rewriting their reports so that they are in a consistent format.

The applications presented cover a rich and diverse array of topics. This is due in part to the fact that the students themselves have very different backgrounds. It is, however, also due to the fact that the students develop an interest in the different capabilities provided by GAs. Most of the chapters in this book focus on the use of GAs for solving engineering optimization problems. There are, however, also chapters that address the use of learning classifier systems and genetic programming. No matter what the particulars of the problem discussed, the intent of this book is to provide examples of imaginative coding schemes and effective fitness functions in hopes that they will encourage readers to develop their own fitness functions and coding schemes, and to demonstrate that GAs should now fit easily into the practicing engineer's toolbox.

SUMMARY

GAs have come a long way since their inception by John Holland at the University of Michigan in the early 1970s. Their history is rich with insightful discoveries by a variety of researchers including many of Holland's students. These search techniques based on natural genetics are no longer just abstract concepts suitable for solving academic problems; they are effective optimization and machine learning tools that can be used by practicing engineers and scientists to solve complex, real-world problems.

At one point, successful GA applications required (and were often driven by) a GA expert who brought to the problem a detailed working knowledge of GA theory. Although a familiarity of how a GA operates is certainly valuable in complex problems, such knowledge is not required to use a GA effectively in most applications. Rather, what is required is a detailed knowledge of the problem domain, access to efficient GA code, and an understanding of how to formulate the problem so that a GA can be employed in its solution. In other words, GAs have advanced to the point where professionals who are comfortable writing computer code can apply a GA to their particular problems of interest; GAs have made their way into the practicing engineer's toolbox.

The goal of this chapter is to lay the foundation for the remainder of the book. Subsequent chapters describe computer projects completed by graduate students in a semester course on GAs offered at the University of Alabama. The complexity of the problems that are presented, and the wide spectrum of application domains, serve to provide readers with an indication of the robust nature of GAs. Additionally, the coding schemes and fitness functions described should provide the reader with examples from which they can draw in their efforts to implement GAs in their own problems.

REFERENCES

[1] Rechenberg, I. (1965, August). *Cybernetic solution path of an experimental problem* (Royal Aircraft Establishment Translation No. 1122, B. F. Toms, Trans.). Farnborough hants: Ministry of Aviation, Royal Aircraft Establishment.

[2] Fogel, L. J., Owens, A. J., & Walsh, M. J. (1965). *Artificial intelligence through simulated evolution.* New York: John Wiley & Sons.

[3] Holland, J. H. (1975). *Adaptation in natural and artificial systems.* Ann Arbor: The University of Michigan Press.

[4] Baur, R. J. (1994). *Genetic algorithms and investment strategies.* New York: John Wiley & Sons.

[5] Davis, L. D. (Ed.). (1991). *The genetic algorithm handbook.* New York: Van Nostrand Reinhold.

[6] Aspnäs, A., Cockcroft, V., & Hekonaho, J. (1996). Use of genetic algorithms to learn ligand recognition concepts: application to the GPCR superfamily. In J. T. Alander (Ed.), *Proceedings of the Second Nordic Workshop on Genetic Algorithms.* 131-150.

[7] Yen, J., Liao, Lee, B., & Liao, J. C. (1996). Using fuzzy logic and a hybrid genetic algorithm for metabolic modeling. *Proceedings of the Fifth IEEE International Conference on Fuzzy Systems* (pp. 220-225). Piscataway, NJ: Institute of Electrical and Electronics Engineers, Inc.

[8] Koza, J. R. (1992). *Genetic programming: On the programming of computers by means of natural selection.* Cambridge: MIT Press.

[9] Schwefel, H.-P. (1975). *Evolutionsstrategie und numerische optimierung.* Doctoral dissertation, Berlin: Technische Universitat.

[10] Schwefel, H.-P. (1977). *Numerische optimierung von computer-modellen mittels der evolutionsstrategie.* Basel: Birkhäuser.

[11] Box, G. E. P. (1957). Evolutionary operation: A method for increasing industrial productivity. *Journal of the Royal Statistical Society,* C6 (2), 81-101.

[12] Martin, F. G., & Cockerham, C. C. (1960). High speed selection studies. In O. Kempthorne (Ed.), *Biometrical genetics.* Amsterdam: Pergamon Press.

[13] Bledsoe, W. W. (1961). The use of biological concepts in the analytical study of systems. Paper presented at ORSA-TIMS National Meeting, San Francisco, CA.

[14] Baricelli, N. A. (1957). Symbiogenetic evolution processes realized by artificial methods. *Methodos,* 9(35-36), 143-182.

[15] Fraser, A. S. (1957a). Simulation of genetic systems by automatic digital computers: I. Introduction. *Australian Journal of Biological Science,* 10, 484-491.

[16] Fraser, A. S. (1957b). Simulation of genetic systems by automatic digital computers: II. Effects of linkage on rates of advance under selection. *Australian Journal of Biological Science,* 10, 492-499.

[17] De Jong, K. A. (1975). An analysis of the behavior of a class of genetic adaptive systems. (Doctoral dissertation, University of Michigan). *Dissertation Abstracts International*, **36**(10), 5140B. (University Microfilms No. 76-9381)

[18] Hollstien, R. B. (1971). Artificial genetic adaptation in computer control systems. (Doctoral dissertation, University of Michigan). *Dissertation Abstracts International*, **32**(3), 1510B. (University Microfilms No. 71-23,773).

[19] Booker, L. B. (1982). Intelligent behavior as an adaptation to the task environment. (Doctoral dissertation, Technical Report No. 243. Ann Arbor: University of Michigan, Logic of Computers Group). *Dissertations Abstracts International*, **43**(2), 469B. (University Microfilms No. 8214966)

[20] Goldberg, D. E. (1983). Computer aided gas pipeline operation using genetic algorithms and rule learning (Doctoral dissertation, University of Michigan). *Dissertation Abstracts International*, **44**(10), 3174B. (University Microfilms No. 8402282)

[21] Goldberg, D. E. (1989). *Genetic Algorithms in Search, Optimization, and Machine Learning*. Reading, MA: Addison-Wesley Publishing Company, Inc.

[22] Davidor, Y. (1991). *Genetic algorithms and robotics: A heuristic strategy for optimization*. Singapore: World Scientific Publishing Co.

[23] Rawlins, G. J. E. (Ed.). (1991). Foundations of genetic algorithms. San Mateo, CA: Morgan Kaufmann Publishers, Inc.

[24] Michalewicz, Z. (1992). *Genetic algorithms + data structures = evolution programs*. Berlin: Springer-Verlag.

[25] Alander, J. T. (1996). *An indexed bibliography of missing entries* (Report Series No. 94-1-MISSING). Vaasa, Finland: University of Vaasa.

CHAPTER 2

IMAGE-CALIBRATION TRANSFORMATION MATRIX SOLUTION USING A GENETIC ALGORITHM

Thomas P. Dickens
The Boeing Company
e-mail: thomas.p.dickens@boeing.com

ABSTRACT

To aid in the quantitative analysis of a pair of simultaneously taken images, one image can be transformed into the image-space of the other. Analysis can then be done by using the same pixel region from both images. Known points of interest are located on both images as a set of reference marker locations. By using the knowledge that the images were taken at the same time, and of the same item of interest, from approximately the same physical location, a simple 4x4 transformation matrix can be used to map the markers from one image to the other. This chapter outlines an approach to develop, and an implementation of, a genetic algorithm to find an acceptable transformation matrix for a given pair of such images. This research resulted in insights for using a GA in a time-constrained environment, techniques in reducing the effective GA search space, and advanced matrix GA operators.

INTRODUCTION

The author works for the Boeing Company in Aerodynamics Computing, specializing in geometry, graphics, and scientific visualization. One assignment involved working with pairs of digital images of a model, I1 and I2, where the images in an image pair are taken with different optical filters to view different characteristics of the model. A set of registration targets is placed upon the model to assist in down-stream image manipulation. This set of targets is located within each image in its own image-space as shown in Figure 2.1.

Currently, an image-calibration program is used which transforms the second image into the "space" of the first image, S1, creating a new image I2'. The purpose of this is that statistical operations can now be easily done on the image pair I1 and I2' as simple pixel-based operations in the 2D image space, and the image space for both images has the same mapping (S1 == S2'). The approach used in the existing image-calibration program is to take the registration targets in image I2, process them through a 4x4 transformation matrix T21, and end up with the targets in the space of I1. Once this matrix is "found", each pixel in the image can then be transformed into the I1 image space as shown in Figure 2.2 (with over sampling and jittering to avoid aliasing artifacts).

0-8493-9801-0/98/$0.00+$.50

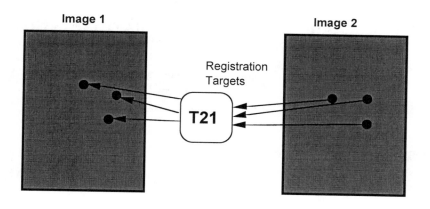

Figure 2.1 - Finding transformation matrix T21 for targets in image 2 to the space of image 1.

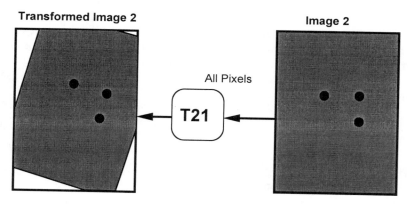

Figure 2.2 - Transforming image 2 through T21 into the space of image 1.

To find the T21 matrix, the image-calibration code calculates an initial guess by calculating the 2-D rotation, scaling, and translation from S2 to S1 for two distant targets. However, this is not a simple 2-D problem, hence the need for a 3-D transformation matrix. The code then cycles on all of the sixteen values in T21, perturbing them slightly, looking for a better "fit" from S2 to S1. The program monitors the sum of the square of the distances from the targets in S2' to the targets to S1. This is analogous to physically jiggling a part to find the best-fit orientation. The program iterates until the fitness is within a given tolerance, or until a maximum number of cycles has been tried. Within this iteration logic an "escape" factor of 0.9995 is used to allow this method to escape from local minimums. The current image-calibration code as shown in Figure 2.3 is used and works well.

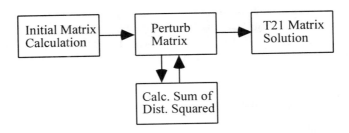

Figure 2.3- Current non-GA block diagram.

INTRODUCTION OF A GENETIC ALGORITHM

In this study, a genetic algorithm (GA) is used to find a suitable transformation matrix T21 for the image transformation problem. As seen in Figure 2.4,
the fitness function needed for the GA is already well-defined in the current
iterative method, as well as the initial matrix calculation algorithm. The goal
with the GA implementation is to achieve a T21 which is as good as, or better
than, the current method, but which converges in less time. The current method
takes an average of 55 seconds to find a suitable mapping when run on an SGI
Indigo2 Extreme. In production use, hundreds of images will be regularly transformed for statistical analysis.

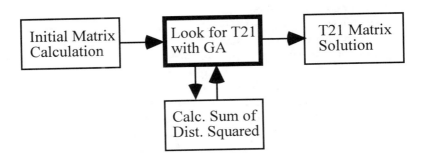

Figure 2.4 - GA-based block diagram.

OTHER METHODS USED

The problem of image-transformation and image-registration occurs in many other fields, such as aerial and satellite photography, medical imaging, industrial imaging, and computer-vision systems. There has been work in these fields with applying genetic algorithms to search for image-to-image and image-to-object mappings.

Dasgupta et al. [1] designed and implemented a structured GA (sGA) for image-to-image registration. The key to their system is a 2-level GA hierarchy with the higher-level genes activating or deactivating sets of lower-level genes, thus achieving a GA which: (1) can maintain diversity over a large number of generations, (2) can achieve a large effect with a single-bit change, and (3) discover potential areas of search space quickly. For their image-to-image registration, they used simple 2-D image vectors at each of the four corners of the image. This method would be too limited for the image-to-image registration used here.

Nagao et al. [2] look at mapping a 3-D object onto a pair of stereo images using a GA. There are many assumptions they make in their system, including camera-to-camera coordination, thus their use of the basic set of x, y, and z rotation and translation parameters is sufficient without needing the additional operators for scaling and perspective. This method uses too many constraints and assumptions to be used for the images in the current problem.

The effort by Wallet et al. [3] looks at the general problem of using a matrix representation with GAs for various problems. Their focus is on a matrix-based crossover operator which operates on a subset of the matrix rather than a sequence of linear-ordered values. For example, a 2x3 sub-set of a 4x4 matrix can be targeted for crossover and operated on as a set. This allows logical groupings of terms within the matrix to be operated on at the same time, thus manipulating high-level concepts within the matrix. This holds true for some cases in a geometry-based 4x4 transformation matrix, such as the upper four values representing the Z rotation. However, the Y rotation uses the non-adjacent matrix locations R11, R13, R31, and R33, and a global scaling uses the locations R11, R22 and R33. The gains achieved by their suggested matrix-grouping could be further expanded upon by defining a set of N crossover templates to use based on the setup of the matrix usage in the particular GA problem. Another approach is to have the GA manipulate the higher-level parameters directly, such as the rotations, scalings, and transformations, then compose the composite matrix when the transformation is required. The work by Wallet et al. has inspired the creation of some advanced matrix operators which are detailed below.

TEST ENVIRONMENT

A set of eight image pairs was identified to use for this initial GA investigation. Given these sixteen images (eight pairs), fifteen of them have been transformed into the space of the sixteenth, and this has been done for all of the sixteen different combinations. Additionally, each of the sixteen images was transformed to itself to ensure that the system can correctly generate an identity matrix. This yielded a total of 256 image-pair transformations to use for testing.

A second group of eight image pairs was then used to validate the results.

GA CODING SCHEME

To scope the problem, the 256 transformation solutions were run using the existing method. These solutions were analyzed to find "typical" ranges for each of the 16 matrix values, which were used to help define the subset of the search-space which is applicable for this problem. Follow-on tests were run and confirmed that this restricted space is reasonable for the problem in general.

Table 2.1 - Survey of 256 initial transformation matrices.

------------------- AVERAGES -------------------			
.99647284237	0.004078072022	0.0000000000	0.0000000000
0.0040780720	0.996472842379	0.0000000000	0.0000000000
.00000000000	0.000000000000	1.0000000000	0.0000000000
.04210090637	8.246609449386	0.0000000000	1.0000000000
------------------- MINIMUMS -------------------			
.93118642603	0.126938044559	0.0000000000	0.0000000000
0.1431280079	0.931186426030	0.0000000000	0.0000000000
.00000000000	0.000000000000	1.0000000000	0.0000000000
178.83712768	161.4008789062	0.0000000000	1.0000000000
------------------- MAXIMUMS -------------------			
.05537870300	0.143128007908	0.0000000000	0.0000000000
.12693804455	1.055378703005	0.0000000000	0.0000000000
.00000000000	0.000000000000	1.0000000000	0.0000000000
86.118011474	28.21191406250	0.0000000000	1.0000000000

The survey data looks at the average as well as the minimum and maximum values for the non-GA run of 256 transformations. Table 2.1 shows the results of applying a single rotation, scaling, then transformation for the initial calculated guess.

Table 2.2 shows the ranges for the final matrix for all 256 cases. Many interesting observations can be made from this data. The upper-left 2x2 section of the matrix represents the rotation about Z, the axis normal to the image, and also includes the X and Y scaling terms [4, 5]. As expected in this problem, these values are well used. The two values on the bottom to the left are the X and Y translation values, and as expected, these have the largest swing to line-up the images. The two values on the top of the rightmost column are perspective, or shearing terms. These compensate for one image being tilted away in relation to the other image. The italicized rows and columns are essentially the values from the identity matrix, which indicate that X and Z rotations, as well as the Z scaling, are not required. The deltas as seen in Table 2.3 for these locations are essentially zero. The GA can perhaps be simplified by not including these seven data values as part of the GA search space. The bottom right value is used as a weighting factor used in the mapping of the combined affine transformations into homogeneous coordinates. The range of this value is quite small, and perhaps it, too, can be left out of the GA search space, if desired.

Table 2.2 - Survey of 256 final transformation matrices.

----------------- AVERAGES -----------------			
.00460495973	0.0062431707	0.0000000000	0.0000023771
0.0027754894	1.0021462372	0.0000000000	0.0000052709
.00000000000	0.0000000000	0.9999999997	0.0000000000
.41538202893	10.227446414	0.0000000000	1.0029214879
----------------- MINIMUMS -----------------			
.98017712897	0.1251674242	*0.0000000000*	0.0000666636
0.1097463339	0.9260208138	*0.0000000000*	0.0000396919
.0000000000	*0.0000000000*	0.9999999971	*0.0000000000*
180.84157368	164.12776121	*0.0000000000*	0.9864702952
----------------- MAXIMUMS -----------------			
.03541033425	0.1476586930	*0.0000000000*	0.0000762130
.10043706060	1.0831681440	*0.0000000000*	0.0000623998
.0000000000	*0.0000000000*	*1.0000000000*	*0.0000000000*
81.211965406	23.139480799	*0.0000000000*	1.02629497331

In Table 2.3, the delta ranges from the initial matrix to the final matrix for all 256 cases are shown. Again, notice the seven values which are essentially zero and are ignored by the GA implementation.

Table 2.3 - Survey of the deltas from the initial to the final transformation matrices.

------------------ AVERAGES ------------------			
0.0081321173	0.00216509825	0.0000000000	0.00000237714
0.0013025826	0.00567339514	0.0000000000	0.00000527097
.00000000000	0.00000000000	0.0000000003	0.00000000000
.62671887743	1.98083696510	0.0000000000	0.00292148791
------------------ MINIMUMS ------------------			
0.0599155533	0.01406627527	0.0000000000	0.00007621304
0.0402972889	0.02787662752	0.0000000000	0.00006239981
.00000000000	0.00000000000	0.0000000000	0.00000000000
0.3963635975	0.67255857052	0.0000000000	0.02629497331
------------------ MAXIMUMS ------------------			
	0.00467318981	0.0000000000	0.00006666364
.03698794842	0.00532152847	0.0000000000	0.00003969191
.00000000000	0.00000000000	0.0000000002	0.00000000000
.45551843191	7.02071723668	0.0000000000	0.01352970472

The delta range covered for each matrix location is shown in Table 2.4.

Table 2.4 - Range covered from delta-minimum to delta-maximum.

------------------ RANGE ------------------			
.08480952415	0.01873946509	0.0000000000	0.00014287669
.07728523735	0.03319815599	0.0000000000	0.00010209172
.00000000000	0.00000000000	0.0000000029	0.00000000000
.85188202946	7.69327580720	0.0000000000	0.03982467803

Table 2.5 - Minimum and maximum value ranges as implemented.

-------------- MINIMUMS --------------			
0.06	0.015	.0	0.000077
0.041	0.028	.0	0.000063
.000	.000	.0	.0
0.40	0.70	.0	0.03
-------------- MAXIMUMS --------------			
.025	.0047	.0	.000067
.037	.0054	.0	.000040
.000	.0000	.0	.0
.5	.1	.0	.014

The actual minimum and maximum delta ranges implemented within the GA are shown in Table 2.5. The range was expanded slightly to ensure all of the desired values fit within the search-space. Additionally, the GA code was implemented with a global matrix scaling factor to further expand this search-space for investigation.

The GA code is used to identify a set of matrix deltas which, when added to the initial matrix, generate the T21 transformation matrix. To represent these delta range values in the GA code, each matrix delta value is represented by an N-bit string. For an N-bit string used at each matrix location, the resulting data coarseness of that value would be:

$$\text{Coarseness}_{i,j} = \frac{\text{range}_{i,j}}{2^N - 1} \qquad (2.1)$$

The data calculated in Table 2.6 shows the effect of increasing the number of bits used in the delta coding. A simple linear mapping from the minimum to the maximum value for each matrix location is used. The concern here is that if N is too small, then a good T21 cannot be found because, as the delta is incremented from one possible value to the next possible value, the requisite value is bypassed.

Table 2.6 - Resulting coarseness calculated for different numbers of bits.

Range	Coarseness						
	N=4	N=8	N=12	N=16	N=20	N=24	N=28
0.031	0.002	8.1E-06	1.9E-09	3.0E-14	2.8E-20	1.7E-27	6.4E-36
0.0197	0.001	5.1E-06	1.2E-09	1.9E-14	1.8E-20	1.0E-27	4.0E-36
0.0001	9.6E-06	3.7E-08	9.1E-12	1.4E-16	1.3E-22	7.9E-30	2.9E-38
0.078	0.005	2.0E-05	4.9E-09	7.6E-14	7.2E-20	4.3E-27	1.6E-35
0.033	0.002	8.7E-06	2.1E-09	3.2E-14	3.1E-20	1.8E-27	6.8E-36
0.0001	6.8E-06	2.6E-08	6.5E-12	1E-16	9.5E-23	5.7-30	2.1E-38
6.9	0.46	0.018	4.4E-07	6.7E-12	6.4E-18	3.8E-25	1.4E-33
7.8	0.52	0.020	4.9E-07	7.6E-12	7.2E-18	4.3E-25	1.6E-33
0.044	0.0029	1.1E-05	2.8E-09	4.2E-14	4.0E-20	2.4E-27	9.0E-36

To allow the needed value to be approximated to an acceptable degree, the number of bits-per-value can be increased. Increasing N, however, will increase the search-space of the problem. Given nine matrix values to search on, for each 1-bit increase in N, there is a 2^9, or 512 times increase in the size of the search space. To allow the GA to converge in a reasonable time, the search space should be kept small.

GA OPERATORS AND TECHNIQUES

In additional to the usual reproduction, crossover, and mutation GA operators [6], additional GA operators and techniques were designed and implemented for this specific problem. The entire set of GA operators implemented for this problem follow:

Crossover

A simple 1-point crossover, as shown in Figure 2.5, is used which swaps the alleles (bit-patterns) in two selected parents at a given bit position. This bit position can be chosen anywhere within the strings, within a delta value as well as between delta values.

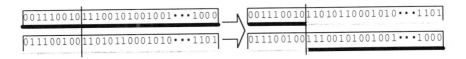

Figure 2.5 – Single point crossover.

Two-Point Crossover

A 2-point crossover, as shown in Figure 2.6, is used which swaps the alleles in two selected parents between two given bit positions. These bit positions can be chosen anywhere within the strings, within a delta value as well as between delta values. For this crossover, the string is considered to wrap back on itself in a ring-like fashion, which results in this 2-point crossover swapping a selection of the ring for two selected parents.

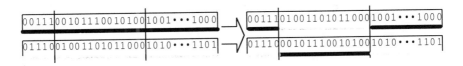

Figure 2.6 – Two-point crossover.

Matrix-Element Crossover

Wallet, et al. [3] inspired the use a matrix-element crossover operator, as seen in Figure 2.7. Their technique of selecting a contiguous sub-region of the 4x4 matrix was expanded to select a number of matrix elements which were related for this specific problem. Templates in the code were designed to define different matrix elements to be used for the eight different matrix-element crossovers possibilities desired. When the matrix-element crossover was selected, one of these templates was then selected and used to accomplish the crossover.

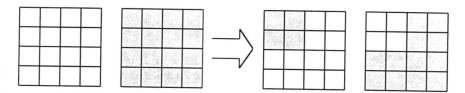

Figure 2.7 - Matrix-element crossover.

The matrix-element crossover swaps selected matrix elements between two selected parents. These swaps are done on full N-bit matrix elements within the strings and does not mix string values within a given matrix element. Figure 2.8 shows the eight matrix-element templates implemented here. The odd templates fit within the Wallet scheme, while the even-numbered templates show the extension to the Wallet scheme. The templates correspond to transformation operators: (1) Z rotation, (2) Z rotation with homogeneous coordinate scaling, (3) X and Y translation, (4) X and Y translation and perspective, (5) X and Y perspective, (6) X and Y scaling, (7) X and Y perspective with homogeneous coordinate scaling, and (8) X, Y, and Z scaling with homogeneous coordinate scaling.

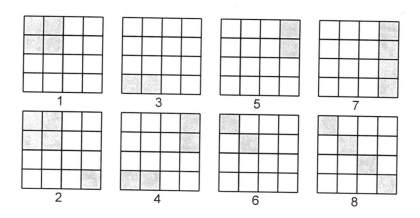

Figure 2.8 - Eight matrix-element crossover templates.

Maintaining the Best Performer

Similar to De Jong's generation-gap idea, the best performer of each generation is copied as-is to the next generation. This guarantees that the solution as found by the GA will not get worse over time. This approach is known as elitist selection.

Varying Mutation Rate

The classic GA implementation uses a fixed mutation-rate. Being concerned with a timely solution, GA stagnation was studied very closely. GA stagnation is defined as the number of generations in which the best performer remains the same, as guaranteed above. One stagnation related solution was to impose an increasing mutation rate (Ms), which is the product of the no-change generation count (NC) and the mutation rate (MR): Ms = NC * MR. This results in ever-increasing chances for mutation to change the best performer in a positive direction.

Comet-Strike Operator

Another GA stagnation operation is, as coined by Robert Dain [7], the Comet-Strike. The analogy is that of a comet striking the earth, which reeks havoc with the ecosystem and destroys all but the most robust of life. Following the comet-strike, new species evolve. The GA equivalent is to maintain the best performer and re-initialize the remaining population. The comet-strike operator is invoked when the no-change generation count (NC) goes above a set threshold and the average/minimum value is less than M (the bulk of the population consists of similar genetic material).

Total-Comet-Strike Operator

A total-comet-strike operator is a re-initialization of the entire population. For this problem, and possibly other real-time GA problems, this operator allows the GA run to result in a better solution at the end of the imposed time-limit.

Local Hill Climbing

A technique borrowed from the current image-calibration code is to take a "good" solution, and to try small permutation on the matrix values to get a better solution. This is done to the best performer as it is copied from one generation to the next. A GA parameter is used to set the maximum number of hill-climbing iterations, resulting in small hill-climbing steps per generation. If needed, a recurring best performer can continue to climb the same hill over many generations.

Gray-Code Mapping

Gray-coding is commonly used with GA implementations due to the 1-bit change in neighboring gray-code values. The option to evaluate the matrix value deltas as gray-coded numbers was provided and investigated.

EFFECT OF CHANGING GA PARAMETERS ON THIS PROBLEM

The GA implementation uses various parameters during a matrix-solution run. The effect of these parameters was studied in depth to access the parameters' influence on the convergence and timeliness of the run. There were some unexpected observations made in this study.

These studies were designed to have the GA run for about 30 seconds maximum. The resulting distance value was tracked as well as the time to run. Five runs at each setting were used to get a statistical spread of the parameter-setting effects. Rather than include pages and pages of multi-line plots, the main points of these studies are summarized. The author can be contacted for access to the detailed analysis.

Maximum Generations (MG) and Population Size (PS)

These two parameters are tightly intertwined in this problem due to the time deadline constraint; an increase in the population size will cause each generation to take longer to run. Population sizes in the range of 10 to 100 individuals were studied. Without accounting for the time factor, larger populations generally had better convergence values for a given number of generations. It was observed that the relation between MG and PS is a linear relation, where PS * MG = 1,000,000 resulted in a maximum-generation stopping time of about 30 seconds. A change in either PS or MG would yield the required value for the other. The population study was re-run with this time-normalized relationship, which showed an advantage in convergence value in a given time period to a population size of about 40 individuals. A PS of 40 requires a MG of 25,000 for 30 seconds. To achieve a 20 second time for a PS of 40, a MG of 25,000 * 2 / 3 = 17,000 (approximately) was used.

Crossover Types

The 1-point, 2-point, and matrix-element crossover types were used with various usage weights. With 100% usage of the 1-point, or 80% or more of the matrix-element crossovers being used, the GA did not converge within the desired time. Various average mixes of these three crossover operators gave good results; there was not a clear mix which outperformed other mixes. The setting of

20%, 40%, 40% for the 1-point, 2-point, and matrix-element crossover operators proved to be a good robust setting.

Number of Bits

The number of bits used (N) for each matrix delta value, and the resulting size of the search space had little effect on the operation of this GA. As expected, small values of N (8 or less) resulted in unacceptable results. Surprisingly, in the range from N=8 to N=64 bits, there was little difference in the performance of the GA in terms of time-to-converge and resulting distance value when finished.

Maintaining the Best Performer

Maintaining the previous generation's best performer in the current generation proved to be a valuable technique. Without this feature, runs were observed which would be converging, then lose the best performer and jump up a bit, continue to converge, then jump again, and so on, as depicted in Figure 2.9. While the average fitness of the population may have been continually decreasing over this entire period, the fact that the GA may be stopped at any given generation and the current best solution is to be used, made this unacceptable in a time-constrained GA.

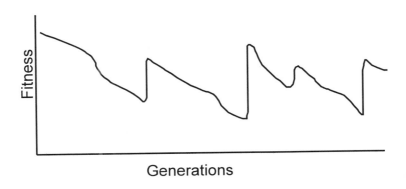

Figure 2.9 - Effect of losing the best-performer on the best-of-generation solution.

Varying Mutation Rate and Comet-Strike Operator

Observation shows that increasingly introducing new genetic material into the population, and also mass introduction of new genetic material into the population, can revitalize the convergence rate in a stagnate run. Without these two mutation-type operations, runs were observed which would stagnate for many hundreds of generations. Given this tendency to favor mutation, a mutation study was undertaken to study the effect of increasing the basic single-bit mutation occurrence. Figure 2.10 shows the results of a series of runs where this mutation rate was increased, and the resulting average fitness for multiple runs is plotted. There is a trend to favor more and more chance of mutation up to a certain point (about 0.002 percent). Additional increases in mutation have a degrading effect on the GA's performance as the system is becoming less of a GA engine and more of a random-search algorithm.

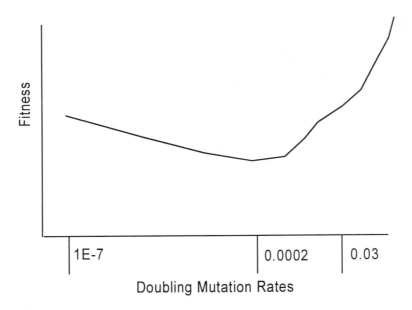

Figure 2.10 - Effect mutation rates on the best-of-generation solution.

Local Hill Climbing

Observation shows that this technique allowed the best performers to slide into a near-by better solution.

Gray Code Mapping

As expected, the use of gray-code mapping had a positive effect on the generation-to-generation performance of the GA, resulting in a 9.5% better convergence value after a given number of generations. However, this is not the whole story. As discovered in implementing the gray-coding capabilities [8], the mapping from a binary integer to gray-code is a simple 1-line equation. However, the mapping from gray-code to a binary integer uses an iterative algorithm which loops on the number of bits (N). This code proved to be rather expensive to execute as was needed for each individual at each generation, and multiple times for the best performers in the local hill-climbing. While 9.5% better in average final value, the gray-code execution time was 2.2 times longer (49 S vs. 22 S for the settings used). With the emphasis on convergence value within a specific time-frame, the use of gray-coding was detrimental and was thus dropped for this problem.

Use of a Matrix Subset

The technique of identifying a subset of the transformation matrix and only searching with nine of the sixteen matrix values proved to both allow acceptable convergence to be achieved as well as providing a time savings. Comparing using all sixteen matrix values to only using nine resulted in a 33% drop in the time-to-run, and a 23% better convergence value. Reducing the scope of the problem based on observed numerical results was beneficial.

GA RESULTS

The sixteen test images were used to test the performance of the GA code. Each image was used as the reference, with the GA mapping the other sixteen images into the image-space of the reference. This provided a sampling of 256 different image mappings, including each image being used as a reference for itself, which should result in a transformation being the identity matrix.

Table 2.7 - Parameters used for the GA.

Number of bits for each delta value	10
Use gray-code encoding	False
Maximum generation	17,000
Population size	40
Percent chance of 1-point crossover	20%
Percent chance of 2-point crossover	40%
Percent chance of matrix-element crossover	40%
Percent chance of mutation	0.0002%
Number of no-change generations before comet-hit	20
Hill-climbing iterations per generation	50

Parameter Notes

The choice of 17,000 generations was chosen to give an average time of about 20 seconds if the GA did not converge.

Test-case Details

After a study of the problem, the originally stated target of 0.1 for the distance value was not required. A value of 0.4 is within the acceptable range for this problem. To provide a fair comparison, the original algorithm was also run with this relaxed distance target; with a resulting average per-run time of 20.0 seconds–with every value below the specified 0.4 figure. It should be noted that the time variance in the original method was very small; each run took the same time within a 10% tolerance.

The 256 test-case transformations were run ten times each, for a total of 2560 GA-based transformations. Table 2.8 shows the results looking at CPU time, GA generations, and the final fitness-function value (the sum of the distances squared). This value is considered by itself, and also as the best and the worst for each of the ten runs for an image pair.

Table 2.8 - Summary of results from the GA run.

Measure	Minimum	Average	Maximum
CPU Time	0.09 S	13.5 S	39.9 S
Generations	62	11,359	17,000
(Sum of Distances)^2	0.143	0.664	2.89
Best of 10 Sums	0.143	0.40	0.99
Worst of 10 Sums	0.40	1.07	2.89

Another look at the resulting data, Figure 2.11, shows the number of runs resulting in ranges of distance values; the tallest bar is in the range from 0.3 to 0.4.

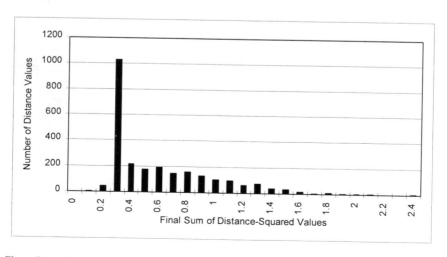

Figure 2.11 - Distance results of the 2560 test-case runs.

Examining the data as a whole, the GA algorithm ran in 9.6 hours versus 14.2 hours for the non-GA method. This is a reduction of 32% in time. The final values achieved did not fare as well, with an average convergence value of 0.664, 66% higher than the specified 0.4 value. As seen in Figure 2.11, over 1,028 runs achieved the 0.4 goal, with 50 of them jumping from above 0.4 to below 0.3 in a single generation–thus 42% of the runs achieved the specified 0.4 goal. The remaining 1,482 runs (58%) did not achieve the 0.4 goal, and are distributed in value ranges in decreasing amounts, with a single run at the largest final value of 2.89.

Looking at select individuals, there were some GA runs which finished in less than 100 mS. Figure 2.12 shows that 17.5% of the runs completed in less than 2.0 seconds and 29% of the runs completed in less than 4.0 seconds. The majority of the runs, 59%, completed in the range of 19.5 to 20.5 seconds. These results indicate that, if "lucky", the GA algorithm will converge in a few seconds, but once the 2.0 second mark is passed, the chance that the GA will converge decreases, until at about 20 seconds the maximum generation count is reached. This timing spread results in an average run time of 13.5 seconds, with two groups: one around 2.0 seconds and a tighter group at 20 seconds.

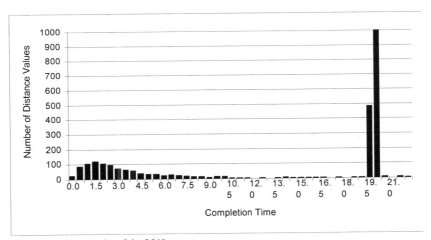

Figure 2.12 - Time results of the 2560 test-case runs.

The correlation between Figures 2.11 and 2.12 is that the 1,078 runs which achieved the distance goal in Figure 2.11 are spread out from 0 to 19 seconds in Figure 2.12, and the 1,482 runs around 20 seconds in Figure 2.12 are spread out above the distance of 0.4 in Figure 2.11. This led to the idea of implementing a Total-Comet-Strike every 4.0 seconds (0.2 * the maximum-generations), to take advantage of a completely fresh gene-pool, thinking that maintaining the best performer overshadowed a new comet-strike generated population and remaining in the area of the local minimum already found. This is essentially running five sequential GAs with MG = MG/5, with the run terminating if any of these sub-runs reaches the convergence value. Preliminary results implementing this additional capability look very promising, with an initial result over thirty-five runs as presented in Table 2.9. All three averages, CPU time, generations, and distance value, have improved from 2.1 to 6.4 percent. The downside is that the maximum distance value has grown to 4.8.

Table 2.9 - Summary of results from the GA run.

Measure	Minimum	Average	Maximum
CPU Time	1.37 S	12.9 S	20.9 S
Generations	1139	10,628	17,000
(Sum of Distances)^2	0.32	0.64	4.8

CONCLUSIONS

A Genetic Algorithm has been successfully implemented to locate an acceptable transformation matrix to map one image into the space of another image. The current state of the GA does find an acceptable solution below the desired target of 0.4 in 42% of the runs in the 2,560-run test-case. Tests with another image-set test-case yields similar results. The desire to better the time using the GA algorithm over the existing iterative approach was achieved for the test case, but the average distance result did not achieve its goal.

Total-Comet-Strike (TCS) Operator

The idea of re-initializing the GA every MG/N generations, by implementing the Total-Comet-Strike operator, was prompted by this effort and looks promising. The next logical addition is to cache all of the final values of the N sub-runs, and if the solution has not converged in the N sub-runs, then either the best of the five should be used and the run terminates, or the best of the N is used and the run continues for another period of time. A time-based or maximum-generation termination needs to be maintained to avoid the rouge case which refuses to converge.

GA Use in a Time-constrained Environment

Revisiting the requirement of achieving a 0.4 distance goal within 20 seconds, on a per-individual case, this goal cannot be guaranteed with the GA approach. However, this requirement needs to be considered for the overall population of runs in the job-set, and if no one run is time critical and the desire is for the entire job-set to complete faster, this GA approach is usable. With the addition of the sub-run TCS technique as described above, the times should be improved even more. As a last resort, for a run which fails to converge, the run can be processed with the existing method with a maximum run-time of twice the targeted 20 seconds (20 for the failed GA, and 20 for the non-GA run). This time could be reduced by avoiding the initialization overhead by carrying both algorithms which share the same front and back ends in the same program (Figures 2.3 and 2.4). This hybrid approach would result in an overall reduction in time for a set of runs, and will also guarantee all runs reach the specified minimum distance value.

Applicability in a Production Environment

With the above mentioned additions to make the algorithm finite in execution time with an acceptable convergence value, this GA-based hybrid method will result in a net time savings without sacrificing the quality of the solution.

Repeatability

A concern with using algorithms which utilize randomness is the repeatability of the solution for the same dataset. In my experience working with engineers, answers, even good answers, which are different for the same set of inputs, are regarded as suspect. With computer-generated randomness as used by the GA, the random-number generator could be seeded with a calculated value derived from the two input images. This would generate the exact same calculated value for the same set of images and parameter settings, which would generate in the same exact result. If the code is to be run on a variety of computers or operating system versions, care should be taken to ensure that the random number generators on the set of computers and operating systems, all behave the same; generating the same set of "random" values. A custom random-number generator could be implemented as part of the GA code, which would run the same across different platforms. The drawback to this is that the random number generator provided with the OS has generally been tuned to perform well on the specific hardware. With the time-critical nature of the requirement, this trade-off would need to be weighed carefully.

Matrix Techniques

The technique of calculating an initial transformation matrix, and then using the GA to find the deltas to this initial matrix, proved to be effective. Identifying the necessary matrix locations and not utilizing the remaining locations also proved to be useful in the performance of this GA implementation. The creation of a sub-matrix crossover which groups matrix values which are related to the matrix usage worked well; it was not the only crossover method required, but the addition of this method aided the algorithm in general.

REFERENCES

[1] Dasgupta, D.; McGregor, D.R. (1992). *Digital image registration using structured genetic algorithms.* Proceedings of the SPIE - The International Society for Optical Engineering, **1766**, 226-34.

[2] Nagao, T.; Agui, T.; Nagahashi, H. (1995). *Fitting three-dimensional models to stereo images using a genetic algorithm.* Proceedings of the SPIE - The International Society for Optical Engineering Conference Title: Proceedings of the SPIE (USA), **2501**, 129-39.

[3] Wallet, B.C.; Marchette, D.J.; Solka, J.L. (1996). *A matrix representation for genetic algorithms.* Proceedings of the SPIE. **2756**, 206-14.

[4] John J. Craig. (1986). *Introduction to Robotics: Mechanics & Control.* Reading, MA: Addison-Wesley Publishing Company.

[5] F. S. Hill Jr. (1990). *Computer Graphics.* New York: Macmillan Publishing Company.

[6] David E. Goldberg. (1989). *Genetic Algorithms in Search, Optimization & Machine Learning.* Reading, MA: Addison-Wesley Publishing Company.

[7] Robert Dain is a student in this class and should have his project included in this publication. He was a fellow Boeing employee, but has since left Boeing to pursue his fortune at Microsoft.

[8] William H. Press et al. (1992). *Numerical Recipes in C: The Art of Scientific Computing, second edition.* Cambridge, MA: Cambridge University Press.

CHAPTER 3

GENETIC ALGORITHMS FOR H$_2$ CONTROLLER SYNTHESIS

Keith J. Nicolosi
Masters Student, Aerospace Engineering
The University of Alabama
Tuscaloosa, Alabama
email: knicolos@eng.ua.edu

ABSTRACT

Modern control theory has evolved over the past decade, with H$_2$, H$_\infty$, and H$_2$/ H$_\infty$ methods gaining recognition in controller synthesis problems. In the event of any plant changes or disturbances, these methods provide a procedure for multivariable controller tuning. There has been substantial research on the development of homotopy algorithms for solving the H$_2$, H$_\infty$, and fixed-order mixed H$_2$/ H$_\infty$ compensator synthesis problems [1]. In order for the homotopy algorithm to perform well, the initial gain matrix used in the algorithm must be very close to the optimal gain matrix, such that the initial gain matrix stabilizes the closed-loop system. Also, the homotopy algorithm requires the use of gradient information, which presents a problem with the discontinuities and vast multimodal, noisy search spaces of controller synthesis. As a remedy for this problem, a genetic algorithm is presented for the synthesis of H$_2$compensators to evaluate the efficiency of the algorithm. The results of this study will be used to determine if a genetic algorithm can achieve the desired performance and robustness characteristics associated with the H$_2$compensator synthesis problem. These characteristics include the requirement of closed-loop stability and minimization of the controller's cost function.

INTRODUCTION

The last decade has witnessed a tremendous increase in the use of vibration isolation systems. These systems have been especially prominent in efforts to protect sensitive devices from vibrations apparent in the environment in which the devices operate. Some examples of the use of these vibration isolation systems include isolating delicate experiments from external vibrations, preventing the transmission of vibrations from rotating machinery in a satellite structure to the scientific sensors, and isolating automobile or aircraft frames from engine vibrations. Current interest in the development of vibration control has motivated research in modern and robust control theory. This is particularly true in those areas related to space structures.

Space shuttle missions have brought about an awareness of the need for autonomous or active vibration isolation for acceleration sensitive payloads.

These active vibration isolation systems will be vital for the success of the microgravity science experiments that will be performed on the International Space Station Alpha. A recent demonstration of active vibration isolation was performed by NASA using the STABLE (Suspension of Transient Acceleration By LEvitation) system. This application for microgravity payloads was the first actively controlled vibration isolation system to be flight tested [2].

Active vibration isolation is also needed for some precision spacecraft that have stringent pointing requirements. These spacecraft are usually flexible, light weight, low stiffness structures, and the stringent pointing requirements result in many closely spaced, lightly damped vibration modes. With the use of an autonomous, robust vibration isolation control system, the spacecraft has the ability not only to reduce the unwanted effects of pointing jitter or excessive motion, but also to re-design, or tune, the control system. These self-tuning spacecraft are able to virtually re-design their control system in the event of any plant changes or disturbances.

The need for this type of autonomous vibration isolation controller tuning was evident with a recent event that occurred to the Hubble Space Telescope [3]. Once the Hubble Space Telescope was in orbit, it began experiencing pointing disturbances due to thermal induced vibration on the solar arrays. A re-design of the control design model was performed with the use of on-orbit data. However, the re-design of the model took important operational time, as well as a substantial investment of manpower, both of which could have been saved with a more efficient on-orbit re-design system. Other events such as deep-space missions stand to benefit from robust re-design systems in the event of disturbances or changes to the spacecraft during its long journey. One approach to acquire this type of self-tuning for multivariable controllers is known as *perturbation compensation*, where the parameters of a dynamic compensator are tuned to account for plant changes and to provide levels of robustness and performance.

This type of self-tuning system would be advantageous for a robust, active vibration isolation controller. A control design methodology that is well suited for this type of controller synthesis is H_2 controller synthesis. The objective of the H_2 problem is to minimize the H_2 norm on the closed loop transfer function from disturbance inputs to performance outputs, while guaranteeing closed-loop stability [1]. The re-design of the dynamic compensator for a H_2 controller is a difficult task. Numerous approaches for H_2 controller synthesis have been developed. However, some methods have several limitations that restrict them from optimal performance. Some of these limitations include long run-times, a dependence on stable, initial guesses which are close to the optimal design, and a dependence on derivative information. To circumvent this difficulty, a genetic algorithm (GA) will be developed for this formulation.

PROBLEM STATEMENT

H_2 optimal control models the control design as a problem of minimizing the H_2 norm on the closed-loop transfer function while utilizing a state or a measurement feedback controller. The H_2 optimal control problem is a deterministic setting of the linear quadratic Gaussian (LQG) control problem. LQG control theory is a powerful design tool which involves a linear model of a plant in a state-space description. The designer assumes properties for disturbances and measurement noise and translates the design specifications into a quadratic performance criterion consisting of state variables and control signal inputs. The designer's objective is to minimize the performance criterion while at the same time, guaranteeing closed-loop stability. This involves the task of solving for the optimal compensator parameters, which are contained in the output feedback gain matrix. The formulation of the H_2 optimal problem proceeds as follows [1].

The generalized plant of a standard control problem is given by

$$\dot{x} = Ax + B_1 w + B_2 u \tag{3.1}$$
$$z = C_1 x + D_{12} u \tag{3.2}$$
$$y = C_2 x + D_{21} w + D_{22} u \tag{3.3}$$

where $x \in R^n$ is the state vector, $w \in R^{n_w}$ is the disturbance vector, $u \in R^{n_u}$ is the control vector, $z \in R^{n_z}$ is the performance vector, and $y \in R^{n_y}$ is the measurement vector. Figure 3.1 illustrates this design framework. The following is assumed.

1) (A, B_1, C_1) is stabilizable and detectable.
2) (A, B_p, C_p) is stabilizable and detectable.
3) D_{12} has full column rank.
4) D_{21} has full row rank.

A general compensator for the system is

$$\dot{x}_c = A_c x_c + B_c y \tag{3.4}$$
$$u = C_c x_c \tag{3.5}$$

where $x_c \in R^{n_c}$ is the state vector of the controller. Closing the loop using negative feedback yields the closed-loop system dynamics

$$\dot{\tilde{x}} = \tilde{A}\tilde{x} + \tilde{B}w$$

(3.6)

$$z = \tilde{C}\tilde{x}$$

(3.7)

where

$$\tilde{x} = \begin{bmatrix} x \\ x_c \end{bmatrix}$$

(3.8)

$$\tilde{A} = \begin{bmatrix} A & -B_2 C_c \\ B_c C_2 & A_c - B_c D_{22} C_c \end{bmatrix}$$

(3.9)

$$\tilde{B} = \begin{bmatrix} B_1 \\ B_c D_{12} \end{bmatrix}$$

(3.10)

$$\tilde{C} = \begin{bmatrix} C_1 & -D_{12} C_c \end{bmatrix}$$

(3.11)

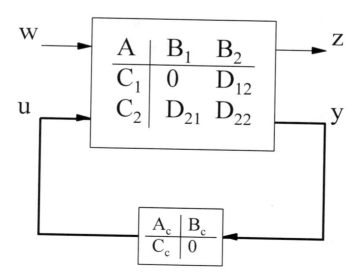

Figure 3.1. Generalized plant with general compensator.

The set of all internally stabilizing compensators is defined as

$$S_c = \{(A_c, B_c, C_c): \tilde{A} \text{ is asymptotically stable}\} \qquad (3.12)$$

For an H_2 problem, the objective is to minimize the H_2 norm on the closed-loop transfer function from disturbance inputs to performance outputs

$$T_{zw} = \tilde{C}(sI - \tilde{A})^{-1}\tilde{B}. \qquad (3.13)$$

If the disturbance is modeled as white noise, the objective is

$$\min_{S_c}\left\{ J(A_c, B_c, C_c) = \lim_{t \to \infty} E\left\{z(t)^T z(t)\right\}\right\} \qquad (3.14)$$

It can be shown that the cost can be expressed as [1]

$$J(A_c, B_c, C_c) = tr\{Q\tilde{B}\tilde{B}^T\} = tr\{P\tilde{C}\tilde{C}^T\}, \qquad (3.15)$$

where

$$\tilde{A}P + P\tilde{A}^T + \tilde{B}\tilde{B}^T = 0 \qquad (3.16)$$
$$\tilde{A}^TQ + Q\tilde{A} + \tilde{C}^T\tilde{C} = 0 \qquad (3.17)$$

P is the controllability grammian of $(\tilde{A}, \tilde{B})$ and Q is the observability grammian of $(\tilde{C}, \tilde{A})$.

In order to obtain the H_2 optimal compensator, the Lagrangian is defined as

$$\mathcal{L}(Q, L, A_c, B_c, C_c) = tr\{Q\tilde{B}\tilde{B}^T + (\tilde{A}^TQ + Q\tilde{A} + \tilde{C}^T\tilde{C})L\} \qquad (3.18)$$

where L is a symmetric matrix of multipliers. Matrix gradients are taken to determine the first-order necessary conditions

$$\frac{\partial \mathcal{L}}{\partial Q} = \frac{\partial \mathcal{L}}{\partial L} = \frac{\partial \mathcal{L}}{\partial A_c} = \frac{\partial \mathcal{L}}{\partial B_c} = \frac{\partial \mathcal{L}}{\partial C_c} = 0 \qquad (3.19)$$

Computation of an H_2 optimal controller for the general compensator requires the simultaneous solution of five coupled equations. This process becomes computationally expensive, and the problem is over-parametrized with such a compensator. To avoid the problem of over-parametrization, either a controller or observer canonical form can be imposed on the compensator structure so that the number of parameters is reduced to its minimal number.

The resulting augmented system defines a static gain output feedback problem where the compensator is represented by a minimal number of free parameters in the design matrix, G. This augmented system is shown in Figure 3.2. The closed-loop system is given by

$$\dot{\overline{x}} = (\overline{A} - \overline{B}_2 G \overline{C}_2)\overline{x} + \overline{B}_1 w = \tilde{A}\overline{x} + \tilde{B}w$$
(3.20)

$$z = (\overline{C}_1 - \overline{D}_{12} G \overline{C}_2)\overline{x} = \tilde{C}\overline{x}$$
(3.21)

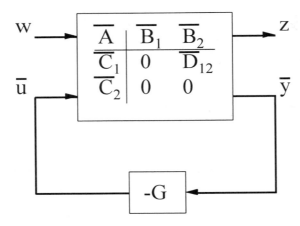

Figure 3.2. Augmented system with compensator in controller canonical form.

The optimal compensator design for this H_2 optimal controller problem is obtained by finding the compensator parameters in the output feedback gain matrix, G, which minimizes the cost function of Equation (3.15). The crux of this chapter is to use a GA to find the optimal controller gain matrix, G, for the H_2 compensator synthesis problem. For this study, the H_2 optimal problem will be solved for a four-disk system (see Figure 3.3).

MOTIVATION FOR USING GENETIC ALGORITHMS

Numerous techniques have been developed to synthesize H_2 controllers, such as Newton's method [4] and homotopy algorithms [1]. However, these methods have several limitations that restrict them from optimal performance. Some of these limitations include long run-times, a dependence on stable, initial guesses which are close to the optimal design, and a dependence on derivative information. GAs are not restricted by these limitations; therefore, a GA will be used to acquire a more robust and perhaps more efficient controller design.

GAs are search algorithms that combine a survival-of-the-fittest approach with a structured, yet random information exchange. This combination provides a balance between the exploration of the search space, and an exploitation of successful solutions. A GA differs from traditional search methods in three main ways [5]:

> 1) GAs use a coding of the parameter set, rather than the parameters themselves;
> 2) GAs simultaneously consider a population of points, rather than a single point;
> 3) GAs use only payoff information, rather than derivatives or other auxiliary information.

GAs require the parameter set of the optimization problem to be coded as a finite string of bits. Since a population of these strings are considered simultaneously, the chance of locating a false peak in a multimodal search space is reduced over methods that go from a single point to a single point. The coding similarities found in each population are used to differentiate good solutions from bad solutions. The use of payoff information gives each string a fitness value, without having to rely on any other auxiliary information. The fitness values for a particular string are obtained from an objective, or fitness function which dictates that string's "goodness" as a solution of the search space. These differences from traditional methods allow a GA to perform well with the discontinuous and vastly multimodal, noisy search spaces of H_2 controller synthesis.

TEST CASE

To verify the validity of using a genetic algorithm for H_2 optimal controller synthesis, a four disk model was studied. The four disk model represents an apparatus developed for the testing of pointing control systems for flexible space structures. A schematic of the four disk system is shown in Figure 3.3. The four disks are rigidly attached to a flexible axial shaft and control torque is applied to selected disks. The angular displacement of the disks are measured. The equations of motion are written as

$$I_1\ddot{\theta}_1 + K_1(\theta_1 - \theta_2) = 0 \tag{3.22}$$

$$I_2\ddot{\theta}_2 + K_2(\theta_2 - \theta_3) - K_1(\theta_1 - \theta_2) = 0 \tag{3.23}$$

$$I_3\ddot{\theta}_3 + K_3(\theta_3 - \theta_4) - K_2(\theta_2 - \theta_3) = 0 \tag{3.24}$$

$$I_4\ddot{\theta}_4 + K_3(\theta_3 - \theta_4) = 0 \tag{3.25}$$

or

$$\overline{M}\ddot{q} + \overline{K}q = \overline{B}u \tag{3.26}$$

where the generalized displacements are the angular displacements of the disks, q^T, and the input vector consists of the moments applied to each disk, u^T. Defining the state vector as $q^T = [q^T, \dot{q}^T]$ results in the state-space formulation

$$\dot{x} = Ax + Bu \tag{3.27}$$

where

$$A = \begin{bmatrix} 0 & I \\ M^{-1}K & 0 \end{bmatrix} \qquad B = \begin{bmatrix} 0 \\ M^{-1}B \end{bmatrix} \tag{3.28}$$

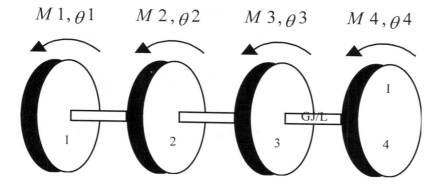

$$M1, \theta1 \qquad M2, \theta2 \qquad M3, \theta3 \qquad M4, \theta4$$

Figure 3.3. Four disk system.

The implementation of the GA to this H_2 optimal problem will be used to study perturbation compensation for the four disk model. Therefore, the initial controller gain matrix will be known and the GA will optimize toward the control gain perturbations. Since the GA will only be solving for the control gain perturbations, the coding scheme can be simplified. The control gain perturbations can be seen as percent variations from the initial control gain matrix, which will reduce the size of the range of each parameter. A "window" of variation for the elements of the matrix are used, such as negative one percent to positive one percent, negative five percent to positive five percent, etc.

The coding scheme that will be used for this multiparameter optimization problem is a simple floating point coding scheme. The G matrix for the four disk model contains sixteen elements, so each parameter will be an individual element of the control gain perturbation matrix, with a range of the "window" pre-defined. Initially, the "window" of variation for the elements of the G matrix are set to a range of minus five percent to plus five percent. With the floating point scheme, each parameter is represented by a floating point number. A study of the "window" size that allows the genetic algorithm to optimize with both robustness and efficiency has been performed.

The objective function used in the genetic algorithm will contain the minimization of the cost function from Equation (3.15),

$$J(A_c, B_c, C_c) = tr\{Q\tilde{B}\tilde{B}^T\} \qquad (3.29)$$

This equation is simply a cost function that involves the trace of the matrix in the brackets, which is the sum of the diagonal terms. Since the objective of the H_2 problem is to minimize the H_2 norm on the closed loop transfer function from disturbance inputs to performance inputs, the cost function will be minimized. This corresponds to minimizing the error associated with the disturbance inputs and performance outputs. Minimizing the trace of the product of matrices in the brackets allows the minimization of this error. In order to insure closed-loop stability, the eigenvalues of the $\tilde{A}$ matrix must all be negative. This knowledge will be incorporated into the objective function by penalizing the fitness, if any, of the eigenvalues are greater than or equal to zero. The fitness function used in the GA is:

$$f = tr\{Q\tilde{B}\tilde{B}^T\} + W \qquad (3.30)$$

where W equals zero if all of the eigenvalues of $\tilde{A}$ are negative, or W equals a positive value, if any, of the eigenvalues of $\tilde{A}$ are greater than or equal to zero. A reproduction scheme is used to select strings with high fitness values for future generations. The type of scheme that was used is called tournament selection, where the string with the lowest fitness value of each round of the simulated tournament is reproduced from a mating pool for the new population. There are popsize rounds to the tournament.

The genetic algorithm was applied to the H_2 optimal controller for three separate cases. The first is used to determine the performance of the genetic algorithm with an initial gain matrix that is deviated within the range of ±5%. The second case incorporates a range of ±10% and the third incorporates an extreme case with a range of ±25%. The objective of the genetic algorithm is to obtain solutions that are near the exact gain matrix values, while requiring closed loop stability. The results of this study help determine the validity of genetic algorithms in this area of controller design.

RESULTS

Using the example of the four disk system, the compensator gain matrix was found, and the minimum cost functional value was 0.0699905. The objective of the GA was to find solutions that contain the matrix elements deviations of the original gain matrix that minimize the cost functional to a value of 0.0699905. For all three cases, the GA was run for 35 generations, using two-point crossover and a mutation operator that changes one of the parameters of a string based on a uniform probability distribution. The results that follow are the fitness values of the best string, averaged from ten different runs, plotted against the number of function evaluations.

For the first case, the exact gain matrix elements were within the range of ±5% different from the initial gain matrix that was used. An initial population of 50 strings was randomly selected, where each parameter represented the amount of deviation between ±5% different than the exact gain matrix elements. The results of this case are displayed in Figure 3.4, showing that the GA obtained a fitness value that was less than one percent larger than the exact minimal cost functional value. The second and third cases contained an initial population of 100 strings with each parameter representing the amount of deviation between ±10% for the second case and ±25% for the third case. The results of the second case are displayed in Figure 3.5. The results show that the GA obtained a fitness value that was less than one percent larger than the exact minimal cost functional value. The results of the third case, shown in Figure 3.6, reveal that the GA did not perform as well as the other cases, obtaining a fitness value that was about six percent larger than the exact minimal cost functional value.

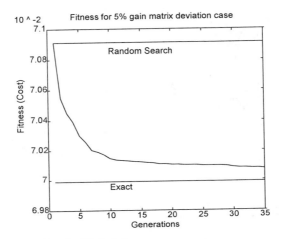

Figure 3.4. GA performance for 5% deviation.

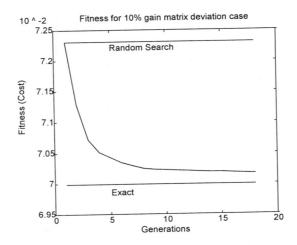

Figure 3.5. GA performance for 10% deviation.

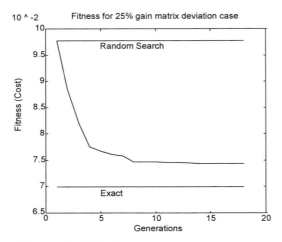

Figure 3.6. GA performance for 25% deviation.

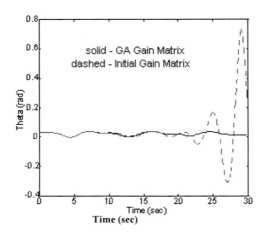

Figure3.7. Disk structure impulse response.

The results of using a GA for H_2 compensator synthesis produced solutions that fulfilled the requirement of closed-loop stability. Figure 3.7 shows the impulse response of the four disk system for the extreme case when the initial gain matrix is deviated within the range of $\pm 25\%$. Output responses are displayed for the initial gain matrix used with the dynamic compensator, and the gain matrix obtained by the GA used with the dynamic compensator. As shown, the response of the system using the dynamic compensator with the initial gain matrix is unstable and diverges quickly. However, the response of the system using the dynamic compensator with the gain matrix obtained from the GA displays a stabilizing effect.

ing the dynamic compensator with the gain matrix obtained from the GA displays a stabilizing effect.

CONCLUSIONS

The results of this study show that the GA led to near-exact solutions, which lowered the value of the cost function, well beyond that which would have been obtained by a random search procedure. The GA obtained fitness values that were less than one-percent of the exact minimal cost functional value for gain matrices that were within the range of $\pm 10\%$, and fitness values that were about six-percent of the exact minimal cost functional value for the extreme case where the gain matrices were on the range of $\pm 25\%$. Also, the requirement of closed loop stability was maintained up to the extreme case of a $\pm 25\%$ deviation of the exact gain matrix. Furthermore, the GA was able to begin with an unstable initial solution, and find a near-optimal stable solution. This is a very important aspect for compensator design since traditional methods have problems with unstable initial conditions. From these studies, the GA has proven to be a valid algorithm to determine the gain matrix of a dynamic compensator for H_2 compensator synthesis.

ACKNOWLEDGMENTS

I would like to express my appreciation to Dr. Michael Freeman and Dr. Charles Karr for their expertise, support, and encouragement. I would also like to thank Dr. Mark Whorton of NASA/Marshall Space Flight Center for his help and expertise during this research project. This research has been supported by the AIAA/Francois-Xavier Bagnoud Foundation and the Department of Aerospace Engineering and Mechanics at the University of Alabama.

REFERENCES

[1]   Whorton, M., Calise, A., and Buschek, H. (1996). Homotopy algorithm for fixed order mixed H_2/H_∞ design. *Journal of Guidance, Control, and Dynamics*, **19**(6, November-December), 1262-1269.

[2]   *Results of the STABLE Microgravity Vibration Isolation Flight Experiment. (1998)*. Internet, http://zaphod.msfc.nasa.gov/~httpser/papers/design/whorton5.html.

[3]   Sharkey, J., Nurre, G., Beals, G., and Nelson. (1992). A chronology of the on-orbit pointing control system changes on the Hubble space telescope and associated pointing improvements. *Proceedings of the 1992 AIAA Guidance, Navigation and Control Conference*, Aug. 10-12, Hilton Head Island, SC.

[4] Seinfeld, D., Haddad, W., Berstein, D., and Nett, C. (1991). H_2/H_∞ controller synthesis: Illustrative numerical results via quasi-Newton methods. *Proceedings of the 1991 American Control Conference, Boston*, MA, June 26-28, 1155-1156.

[5] Goldberg, David E. (1989). *Genetic algorithms in search, optimization, and machine learning*. Reading, MA: Addison-Wesley.

CHAPTER 4

SOFTWARE TEST DATA GENERATION FROM A GENETIC ALGORITHM

Ken Borgelt
Corporate Software Center, US
Motorola Inc.
1301 E. Algonquin Road
Schaumburg, IL 60196
e-mail: ckb009@email.mot.com

ABSTRACT

The work presented in this chapter investigates how a genetic algorithm may be used to produce software unit test data for use in structural testing. A unit test attempts to execute each line of source code in a software module at least once. It is typically performed using a source level debugger after a software module is written or modified. The particular path taken through the module (and there may be many) is determined by the value of any input data as well as data returned from function calls or other sources. Prior art focuses on configuring a genetic algorithm to identify test data executing a specific path, block of code, or control statement. The approach considered here differs in that its goal is to optimize a population of test data which as a whole provides the desired module coverage. This chapter documents how varying population sizes and a new fitness function influence the formation of a suitable set of test data for a simple software program. The fitness function described here provides improvement over the prior art. The performance gains observed result from making the genetic algorithm aware of the structure of the software module being tested. This work is the first step toward the creation of a software development utility to automatically produce unit test data.

INTRODUCTION

Computer software will continue to increase in size and capability as long as customers press for new features and manufacturers keep designing larger storage devices. This is colloquially known as Parkinson's Law of Data. Software is also being used more often in mission critical applications and embedded products where error free operation is paramount. At the same time, companies like Motorola are pressing developers to take significant steps to reduce a product's time-to-market.

Motorola's Land Mobile Products Sector produces two-way radio based communications systems for municipalities, governments, and utilities. Some land mobile systems, such as those providing state wide coverage for Michigan and

Florida, are quite large and complex. Systems are comprised of numerous specialized pieces of equipment. Each of these is an embedded system with at least one microprocessor running software designed to perform some specialized task in the system, such as system control, resource assignment, audio routing, or radio transmit and receive processing. The Repeater Software group produces software for one of these system boxes, an embedded multiprocessor base station. The software used in this product has been developed incrementally over the past seven years by a large number of engineers.

Software that has been maintained over time is commonly called "legacy" code. Adding new features to legacy code is usually the most cost effective way to add new functionality to an older product. This practice is not always painless because it is impossible to anticipate design decisions for features scheduled for development in the distant future. Poor choices early in the development cycle can be difficult to rectify because tight development schedules and overlapping development phases can limit redesign efforts. The result is a poorly designed feature marked by overly complex and error prone software modules. Unit test plans for such modules tend not to be maintained because of their high complexity. Also, developers sometimes do not execute test plans for such modules even when they are available. Instead, developers will test only the functionality they have changed, increasing the risk of not catching a serious problem in the software.

The worst scenario is when a developer modifies a module that is complex enough to impair understanding. The addition of some feature to such a module must be performed so that the existing features continue to work. If the developer breaks one of the older legacy features and does not perform a full unit test, he or she may not find the flaw until system test or later. The worst situation is a bug in software released to customers.

UNIT TESTING

Testing occurs at many steps in the software development lifecycle. The purpose of testing is to identify program flaws. Software is typically broken down into numerous modules, functions or subroutines. The process of testing these individual portions of a program is termed unit testing. Other forms of testing occur throughout the development process. In addition to testing individual software modules, testing also takes place during module integration, at the box and system level, and during field trials. This chapter addresses the production of test data for testing at the software unit or module level. Performing the test reaffirms the module's adherence to the requirements and helps uncover flaws. Test plans written for individual software modules are called unit test plans and consist of multiple calls passing input data to the module under test (MUT). For each set of input data, the test plan indicates what the module's response should be. Unit test plans are typically generated manually by the developer when designing a new module.

Some of the more popular unit testing techniques are equivalence partitioning, structural testing, constraint based testing, and functional testing. Module testing without detailed knowledge of the source code is called black box testing. Black box testing proceeds with the tester being provided no more than a description of the module under test and allowed parameter values. White box (or structural testing) depends on the tester having source code available to help identify test data likely to draw out design flaws. Functional testing and equivalence partitioning are black box testing techniques. Structural and constraint based testing are white box testing techniques. The generation of test data is a fundamental problem of all testing techniques.

In general, any form of software testing tries to limit the number of test cases run. It is simply not economical (or even feasible) to run through all possible permutations of program control and data flow because the number of possible paths or threads of execution through a module can quickly become unmanageable. Structural testing attempts to execute a set of test cases which evaluates every line of code at least once, comparing the results to those expected.

The goal of the current effort is twofold: (1) to investigate the applicability of genetic algorithms to automate unit test data generation, and (2) to identify factors which improve the performance of the genetic algorithm in generating test data. This is the first step in designing a utility to automatically generate unit test data.

Much work remains in the development of this utility. A C language parser is needed to analyze and instrument the source code and then set up the genetic algorithm. The utility would be invoked by adding a new .ut rule to a development project's Makefile. The Unix make utility would instruct the parser to analyze and instrument a copy of the indicated source code file. The parser would then tailor the genetic algorithm's source code to the module under test. The instrumented copy of the module under test and the genetic algorithm's source code would then be compiled and executed to produce a list of test data.

If the MUT included a description of its software requirements, delimited by a special token the parser was designed to detect, the requirements could be extracted and saved by the parser. A suitable test plan could then be constructed by concatenating the saved requirements and driver code calling the module under test with the newly produced test data. The tester is responsible for verifying that actual performance matches the listed requirements. As described, such a utility would produce unit test data on demand and provide the following benefits:

- Eliminate the need for the storage and maintenance of unit test plans
- Reduce cycle time by automating the production of test data

- Eliminate traceability problems when multiple versions of a source module exist

PRIOR ART

Genetic algorithms (GAs) have previously been used to generate both hardware and software test data. Several authors have recently begun to apply genetic algorithms to the unit test process.

Sthamer et al. [1] researched the use of genetic algorithms to produce test data for programs written in the ADA computer language. Their work focused on evaluating a GA's performance on three sample programs. Various codings and fitness functions were evaluated for these sample programs. The goal of this research was to identify coding and fitness functions which could be used to select test data more likely to identify program faults. This was accomplished by configuring the genetic algorithm to reward those test cases which come closest to making the module's decision logic change the thread of execution. If the set of all test data which causes a specific block of code within the module to be executed is thought of as the solution space, the sets of test data rewarded by Sthamer's fitness function are boundary data and are thus more likely to detect faults in the program's control logic.

Xanthakis et al. [2] investigated the automatic generation of unit test data for Pascal programs by analyzing source code and producing appropriate test data. The tool they produced analyzes a source module for constraints and then sets up a genetic algorithm to target a specific path through the module. This process is repeated for each path requiring verification.

Watkins [3] based her work on the earlier work by Xanthakis [2]. Watkins' paper compares the efficiency of test data generation from a genetic algorithm to randomly generated data, showing the GA based technique to be much more effective. Two test programs were instrumented: trityp-easy and trityp-hard. These programs take three integer parameters and (1) determine if they form a triangle, and (2) identify the type of triangle formed. The trityp-hard program also checks for a right triangle. Test data was obtained from two sources, a random number generator and a genetic algorithm. For the trityp-easy program the GA only needed to examine 0.8% of the search space to find parameter values traversing the ten possible paths of execution through the trityp-easy program. Multiple GA runs were made. The results of each run (the paths traversed) were recorded and used to determine when 100% coverage was obtained. Random number generation covered 5% of the search space before traversing all possible program paths.

Watkins' genetic algorithm used a small population size (twenty) and a fitness function which rewarded test data inversely proportional to the number of times that the resulting path through a module was executed. Here, path is defined as the set of linear code sequence and jumps, or code blocks, which were traversed. This is the same as mapping the nodes traversed in a control graph of the module under test. This fitness function rewarded data sets which traversed new code segments, but it did not actively seek them out.

This chapter most closely builds on Watkins' work [3], though there are several major differences. Watkins has shown that a genetic algorithm is much more efficient than random selection at producing test data. Total coverage was achieved only with multiple runs of the genetic algorithm. The work described in this chapter seeks to produce a stable population of test cases in one run of a genetic algorithm.

Genetic algorithms are generally used to find *one* near optimal solution to some problem. The desired solution in this application is not one set of test data (consisting of parameters to the module under test), but rather all sets of test data which, as a whole, provide 100% module line coverage. This type of problem can be thought of as the formation of niches occupied by sets of test data executing a common path through the module under test. The concept of niche formation will be investigated in future work. All of the papers referenced above configured their genetic algorithms to use small population sizes, usually less than 50. The work presented here studied the effect of varying population size as a function of module complexity. The rationale is this: given the fact that there are usually multiple paths through a software module, test data formation is hindered when additional offspring for one highly fit path come at the expense of those traversing some other path. The solution is simply to increase the population size such that the probability of eliminating all test data for any path is remote.

Another major difference is the fitness calculation. Instead of basing fitness on the inverse of the number of times a path through the module was traversed ("inverse path probability", termed IPP in this chapter), fitness is based on the probability of traversing the last block of code in the path through the module. This new fitness function emphasizes or rewards the traversal of unique terminal code blocks, producing a lower fitness when a path executes a block shared with other paths. Another difference is a fitness "bonus" given to test data that execute nodes on the control graph having untraversed children. The bonus helps the GA search for hard to reach code fragments.

Several of the authors reviewed the creation of a utility to generate unit test data with genetic algorithms. However, none of these authors proposed a large population, single run approach. Likewise, none of the authors disclosed how their fitness function operated.

Finally, Watkins' work does a very good job of demonstrating that genetic algorithms are much better than random selection at producing test data. The fitness function examined in this chapter is similar to Watkins', but is unique enough, and provides sufficient performance gains, to stand on its own.   This being said, the author believes the work described in this chapter to be novel and worthy of investigation.

PROBLEM STATEMENT

The work presented here used a modified version of Goldberg's [4] Simple Genetic Algorithm (SGA) to generate test data for one of the test programs used in Watkins' work [3].   The program, trityp-easy, shown in Figure 4.1, is composed of multiple nested if-then-else constructs.

```
1    #include "instrmnt.h"
2
3
4
5
6
7    void trityp(int x, int y, int z)
8    {
9    int triangle_type ;
10
11   INSTRUMENT_CODE;    12
13   if ( x <= 0 || y <= 0 || z <= 0 )
14     {
15     INSTRUMENT_CODE;    16     triangle_type = 4;
17     }
18   else
19   NT_CODE
20
21   triangle_type = 0 ;
22     if (x == y)
23       {
24       INSTRUMENT_CODE;
25       triangle_type += 1 ;   26        }
27
28     if (x == z)
29       {
30       INSTRUMENT_CODE;
31       triangle_type += 2 ;
32       }
33
34     if (y == z)
35       {
36       INSTRUMENT_CODE;
37       triangle_type += 3 ;
```

```
38      }   39
40      if (triangle_type == 0)
41        {
42        INSTRUMENT_CODE;
43        if ( x+y <= z || y+z <= x || x+z <= y)
44          {
45          INSTRUMENT_CODE;
46          triangle_type = 4 ;
47          }
48        else
49          {
50          INSTRUMENT_CODE;
51          triangle_type = 1 ;
52          }
53        }
54      else
55      if (triangle_type > 3)
56        {
57        INSTRUMENT_CODE;
58        triangle_type = 3 ;
59        }
60      else
61      if (triangle_type == 1 && x+y > z)
62        {
63        INSTRUMENT_CODE;
64        triangle_type = 2 ;
65        }
66      else
67      if (triangle_type == 2 && x+z > y)
68        {
69        INSTRUMENT_CODE;
70        triangle_type = 2 ;
71        }
72      else
73      if (triangle_type == 3 && y+z > x)
74        {
75        INSTRUMENT_CODE;
76        triangle_type = 2 ;
77        }
78      else
79        {
80        INSTRUMENT_CODE;
81        triangle_type = 4 ;
82        }
83      }
84
```

```
85    if (triangle_type == 4)
86      {
87      /* not a triangle */
88      INSTRUMENT_CODE;
89      printf ("not a triangle\n") ;
90      }
91    else
92    if (triangle_type == 3)
93      {
94      /* 3 sides equal */
95      INSTRUMENT_CODE;
96      printf ("equilateral\n") ;
97      }
98    else
99    if (triangle_type == 2)
100     {
101     /* 2 sides equal */
102     INSTRUMENT_CODE;
103     printf ("isosceles\n" ) ;
104     }
105   else
106     {
107     /* no sides equal */
108     INSTRUMENT_CODE;
109     printf ( "scalene\n" ) ;
110     }
111   }
```

Figure 4.1 – The trityp-easy program.

The difficulty lies in finding test data which executes each line of source code within the MUT. Here test coverage was measured by instrumenting each block of code between control statements. Viewing trityp-easy as a series of code blocks (consisting of one or more statements) punctuated by control instructions helps visualize the flow of execution and is called the control graph. The code blocks in trityp-easy are delimited by opening and closing braces such as found on lines 14 and 17 in Figure 4.1, and represent nodes in the control graph.

Figure 4.2 depicts a portion of the control graph for trityp-easy up through line number 53. The blocks of sequentially executed code become the nodes of the graph. Control instructions form the edges. Measuring test coverage is simply keeping track of which edges and nodes were traversed.

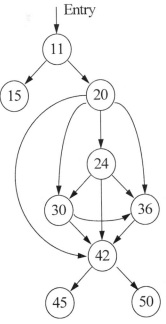

This is a partial control graph for the trityp-easy program up to line number 53. Each circle or node on the control graph represents a block of code. Each number in a circle corresponds to the line number of the 'INSTRUMENT_CODE' macro for that block.

Each directed line segment or edge represents control logic. Note that edges to nodes past 50 are not shown for clarity.

Figure 4.2 – Partial control graph of trityp-easy.

McCabe [5] applied concepts from graph theory to control graphs to define his well known cyclometric complexity metric. This metric provides a measure of the number of linearly independent paths through a software module. The trityp-easy program has a cyclometric complexity of 21. The work described in this chapter varied population size as a multiple of the cyclometric complexity.

Source code was instrumented by adding a function call to each block of code forming a node in the control graph. In Figure 4.1, the macro "INSTRU-MENT_CODE" expands to a function call passing __LINE__ as a parameter. The line number macro is replaced with the source file line number by the C preprocessor. The line number is cached for later use when the MUT executes one of these statements. Edge and node traversal counts are updated with the cached path information after the MUT has been completely traversed by a set of test data. The trityp-easy program has eighteen distinct blocks of code. Each is marked by the 'INSTRUMENT_CODE' preprocessor macro.

SETTING UP THE GENETIC ALGORITHM

As previously indicated, the current effort utilized Goldberg's SGA [4], though some modifications were made to streamline the process of gathering data. Specifically, those portions of the program soliciting operator input were replaced by hardcoded parameter values. All graphical data presented in this chapter were obtained from ensemble averages of 200 runs. Each run lasted 50 generations. Table 4.1 indicates the values of other pertinent GA parameters.

Table 4.1 – SGA parameters

Parameter	Setting
Number of Runs	200
Chromosome Length	24 bits
Number of Generations	50
Probability of Crossover	1.00
Type of Crossover	Two Point
Probability of Mutation	0.005
Scaling	Not Used
Population Size	Varied

Various scaling methods were investigated but were found to actually hinder the formation of stable sub-populations of test data. This is because scaling operators are designed to amplify small differences between individuals, which is a form of positive feedback. The resulting positive feedback is used in standard GA based optimizations to provide good performers with exponentially more offspring. In this application, fitness is initially used to grant more offspring to good performers. However, as a sub-population grows, fitness will decrease, limiting its size and providing negative feedback. This limits uncontrolled reproduction of the descendants of a highly fit ancestor.

CODING

The trityp-easy program used in this work takes three 32-bit integer parameters and determines what type of triangle (if any) they form. The coding used in this work is rather simple. Each 24-bit chromosome is split into three fields or alleles of eight bits each. For trityp-easy, the process of decoding a chromosome into parameters involved a simple conversion of an 8-bit allele into a 32-bit number. This coding was chosen because it was simple and allowed the author to focus on evaluating the fitness function. This coding also decreased the size of the search space, making this problem fairly easy for the GA.

Since the goal of this effort is to eventually produce a usable (automatic) tool, it is not desirable to manually produce a coding for each module evaluated. It would be very efficient to simply code complete function parameters into the chromosome. The drawback of this approach is the large size of the resulting chromosomes. Larger chromosomes take longer to process and evolve. The success of the proposed utility will be dependent on resolving this issue.

FITNESS FUNCTION

This chapter investigates two fitness functions. The first is described in Watkins [3] and is simply the reciprocal of the number of times a given path was executed. Throughout this chapter, the author has termed Watkins' method the Inverse Path Probability (IPP) algorithm to help the reader keep track of the two fitness functions evaluated. The other fitness function is derived from the inverse probability of traversing the last code block in a path and is termed the Last Block Traversal Probability with Bonus, or LBTPB. Most of the work presented compares the performance of the IPP and LBTPB algorithms.

An early alternative to the LBTPB algorithm was derived from calculating traversal probabilities for all edges in the control graph, which trace out the path of execution through the MUT. This produced fitness ratings similar to the probability of executing the last block of code in a given path. The LBTPB algorithm works well for trityp-easy because all paths will terminate with node 88, 95, 102, or 108. The LBTPB algorithm would fare worse if trityp-easy only had one common terminal code block while traversing unique portions of the module earlier in the MUT. This is not as serious an issue as it may seem because the LBTPB algorithm should be capable of being extended to reward unique nodes all along the path.

The path taken through the MUT represents the phenotype of that particular set of test data. The control graph was loaded into a graph data structure implemented as an adjacency list when the GA was initialized. (In the proposed utility, the parser would analyze the MUT and provide control graph initialization information to the genetic algorithm.) Counters in the control graph data structure were used to keep track of how many times each node and edge was traversed. Fitness was calculated after all individual sets of test data making up the population were evaluated.

The relationship between nodes of a graph is often described using the terms of parent and child. Parent nodes have directed edges to child nodes. Additional fitness was given to those sets of test data traversing a parent node having an untraversed child. This fitness bonus helps the genetic algorithm to search for unexplored sections of the MUT. Runs of the genetic algorithm using LBTPB fitness started out with a complete control graph, providing the knowledge needed to provide the bonus. Runs evaluating the IPP fitness algorithm built up the control graph as new nodes were discovered. The IPP algorithm was not made aware of the structure of the MUT and provided no bonus.

The following example illustrates how fitness was awarded. Figure 4.3 shows a sample code fragment and its control graph. The numbers next to the control graph's edges represent traversal counts for that edge. In this example, test cases executing node 16 have not yet been found, making its traversal count zero. However, paths containing node 12, which is the parent to node 16, will receive the fitness bonus. A bonus value of 5 was used in these calculations. The following examples show how the LBTPB fitness may be calculated for the path 1-6-28 and a population size of 16.

The LBTPB algorithm yields a fitness of 2 for the path 1-6-28. The last node in the path (28) was traversed eight times.

$$\text{LBTPB fitness} = \left(\frac{\frac{1}{8}}{16} \right) = \left(\frac{16}{8} \right) = 2$$

The IPP fitness algorithm yields a much higher fitness of 8 for the same path.

$$\text{IPP fitness} = \left(\frac{\frac{1}{2}}{16} \right) = \left(\frac{16}{2} \right) = 8$$

The IPP fitness algorithm yields a much higher fitness of 8 for the same path.

$$\text{IPP fitness} = \left(\frac{\frac{1}{2}}{16} \right) = \left(\frac{16}{2} \right) = 8$$

Thus, for the same path, the IPP algorithm provides a high fitness of 8 while the LBTPB algorithm provides a low fitness of 2. This is an important trait of the LBTPB algorithm: it does not reward paths executing common sections of code. In this example, node 28 is shared among four paths: 1-6-28, 1-12-28, 1-12-16-28, and 1-22-28. For these four paths, fitness should be based on traversal counts of nodes 6, 12, 16, and 22. This matches well with the goal of this chapter – producing a set of test data executing every line of code at least once, without spending the tester's time on redundant tests. Note that the LBTPB algorithm only looks at the last node. Future work will extend this fitness algorithm to reward traversal of unique nodes throughout the control graph, not just the terminal nodes.

The two algorithms provide similar fitness scores for the path 1-12 (after LBTPB awards its bonus). Since node 12 is a parent to the untraversed node 16, paths including node 12 (paths 1-12 and 1-12-28) receive a fitness bonus. For path 1-12 the LBTPB fitness is calculated as:

$$\text{LBTPB fitness} = \left(\frac{\frac{1}{5}}{16} + 5 \right) = \left(\frac{16}{5} + 5 \right) = 8.2$$

The fitness bonus helps the genetic algorithm find untraversed sections of the module under test because descendants of test cases (GA individuals) which include such parent nodes have the best chance of finding untraversed child nodes. Table 4.2 summarizes fitness information for all paths in Figure 4.3.

```
1   INSTRUMENT_CODE;
2   do some initial processing...
3
4   if (A is TRUE)
5     {
6     INSTRUMENT_CODE;
7     process some input...
8     }
9   else
10  if (B is TRUE)
11    {
12    INSTRUMENT_CODE;
13    process other input...
14    if (C is TRUE)
15      {
16      INSTRUMENT_CODE;
      process this hard to reach code...
18      }
19    }
20  else
21    {
22    INSTRUMENT_CODE;
23    process something else...
24    }
25  if (D is TRUE)
26
27    {
```

28	INSTRUMENT_CODE;
29	do some final processing...
30	}

Figure 4.3a -- Code example for fitness calculation.

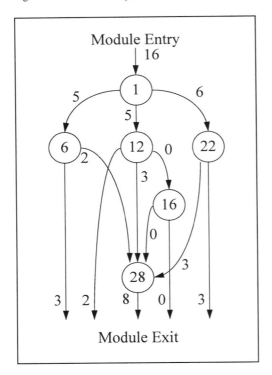

Figure 4.3b -- Control graph for fitness calculation.

Table 4.2 – Comparison of IPP and LBTPB fitness.

Path	#	IPP	LBTPB Before Bonus	Bonus	Total LBTPB
1-6	3	5.33	3.2	0	3.2
1-6-28	2	8.0	2.0	0	2.0
1-12	2	8.0	3.2	5	8.2
1-12-28	3	5.33	2.0	5	7.0
1-12-16	0	N/A	N/A	N/A	N/A
1-12-16-28	0	N/A	N/A	N/A	N/A
1-22	3	5.33	2.67	0	2.67
1-22-28	3	5.33	2.0	0	2.0

POPULATION SIZE

The goal of the current effort is to find all code blocks within the MUT by producing stable sub-populations of test cases traversing as many paths as possible through the module under test. The paths are analogous to the phenotype of an organism in a natural system. Each sub-population forms a niche. The test cases occupying a niche all exhibit the same phenotype, forming a species. In future work, speciation could be promoted using a marriage restriction operator.

It was mentioned earlier that the number of distinct paths through a module can quickly become unmanageable, especially if loops are present within the MUT. Though this issue did not arise while testing trityp-easy (it does not contain any loops), a method of handling paths with loops is needed. The easiest method might be to ignore multiple iterations through the loop, counting the nodes within the loop only once. However, one significant issue remains after this discussion of forming stable sub-populations. How large should these sub-populations be so that they exhibit sufficient genetic diversity and are insensitive to periodically losing a few members?

A multiplier of some measurable software attribute quickly comes to mind. The number of distinct paths through a software module (as noted earlier) can be infinite. Other choices are the number of code blocks (nodes in the control graph) and complexity. Here, McCabe's [5] cyclometric complexity metric was chosen for use as a multiplier. Complexity is a good choice if the proposed utility's parser were able to reduce and instrument complex logic expressions. In the current effort, population size was varied from four to fourteen times trityp-easy's cyclometric complexity in an effort to find a suitable multiplier value. Population size is the product of complexity and the population multiplier.

RESULTS

The IPP and the LBTPB fitness algorithms were compared side by side in multiple runs of the Simple Genetic Algorithm (SGA) developed by David Goldberg [6]. Population size was varied as a fixed multiple of the test module's cyclometric complexity. The test module was the "trityp-easy" program referenced in Watkins [3]. Trityp-easy has a cyclometric complexity of 21 and contains eighteen code blocks. Each run of the GA was initialized with a unique random seed and lasted for 50 generations. Each data point referenced here represents an ensemble average of 200 runs.

Table 4.3 shows the generation number when the two fitness algorithms converged to the eighteen possible code blocks in the trityp-easy program. Both algorithms had difficulty finding all eighteen for low population multipliers. The LBTPB fitness algorithm converged by generation 33 with a multiplier of 8. Increasing the multiplier value advanced the point of convergence. The IPP fitness algorithm was much slower to converge than the LBTPB algorithm, generally converging twenty generations later. Differences between the two fitness algorithms were less pronounced at larger multiplier values. This is because large populations do not adequately challenge the GA. Large initial populations usually contain enough diversity to allow the GA to rapidly traverse all eighteen nodes. However, larger population sizes do consume additional computational resources.

Table 4.3 – Generation of convergence vs. population multiplier.

Multiplier	LBTPB	IPP
4	50+	50+
6	50+	50+
8	33	50+
10	32	50+
12	22	41
14	28	31

Figure 4.4 depicts the convergence of the LBTPB and IPP fitness algorithms when using a population multiplier of 10. The LBTPB algorithm (solid line in Figure 4.4) gave a bonus of 5 fitness points to data sets which traversed a node with at least one untraversed child node.

By providing a bonus, the GA encourages the formation of test data sets for untraversed code blocks. These highly fit individuals cross with other highly fit individuals in the hope of generating test data which are capable of traversing new sections of the module under test. This is what helps the GA converge earlier as shown in Table 4.3 above.

Lower bonus values were less effective, while larger bonus values tended to hinder convergence. The GA is forced to continually rediscover and lose blocks of code when individuals with extremely large fitness values monopolize reproduction at the expense of others.

Figure 4.4 shows convergence of the IPP and LBTPB fitness algorithms. Convergence of the IPP fitness algorithm takes much longer than the LBTPB algorithm. Convergence is shown in greater detail in Figure 4.5, which presents results for the two algorithms with population multipliers of 10 and 12.

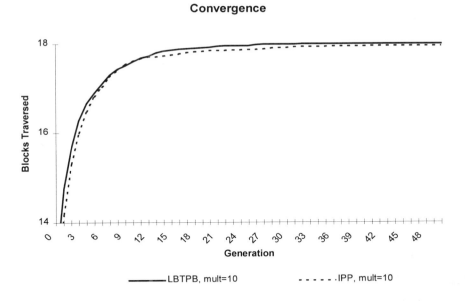

Figure 4.4 – Convergence of LBTPB and IPP fitness algorithms to optimal (18).

Figure 4.5 shows that detection of the remaining code blocks by the two fitness functions stalls momentarily near the optimal value of 18. The LBTPB algorithm clearly performs better than the IPP. Since each trace is actually an average of 200 runs, the plateaus seen near the top indicate that some of the runs have the GA searching for the last few remaining undiscovered sections of the program. This is to be expected from the IPP algorithm because it does not distinguish between unique or common blocks of code when calculating fitness. That these plateaus occurs for the LBTPB algorithm is then somewhat perplexing until we remember that the LBTPB fitness algorithm bases fitness on unique *terminal* blocks of code. As mentioned earlier, the LBTPB fitness algorithm is really just a special case of a fitness algorithm which rewards traversal of unique nodes found along the entire path through the control graph. This more generic form of the LBTPB fitness algorithm is currently under development.

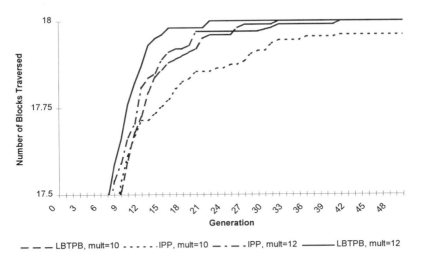

Figure 4.5 – Block coverage near the optimum.

A mutation rate of 0.005 was beneficial. Larger rates were disruptive, while lower rates were less effective. The probability of crossover was set to 1.0, meaning that selected individuals always crossed. This also helped achieve convergence sooner.

The GA was highly susceptible to selection noise. Meaningful results were only obtained after using stochastic remainder sampling and multipoint cross-over (two points over a 24-bit chromosome).

Figure 4.6 is a histogram depicting the number of times each of the eighteen possible blocks of code was executed by both fitness algorithms at generation 50. The data used to produce Figure 4.6 was obtained from GA runs with a population multiplier of 14. This allowed both fitness algorithms to converge well before generation 50. With this large multiplier value both algorithms per-form similarly in terms of speed of convergence. However, the histogram shows that the IPP fitness algorithm executed nodes 80 and 88 (columns 14 and 15 in Figure 4.6) more frequently than the LBTPB algorithm. Node 80 is the block of code entered by the default else statement on line 78 (Figure 4.1). Node 88 is a block of code (beginning on line 86) executed when the function's input parameters do not form a triangle. The LBTPB algorithm produces more sets of test data forming equilateral, isosceles, and scalene triangles (nodes 95, 102, and 108). The LBTPB algorithm distributes test cases more evenly over the four possible outputs (not a triangle, equilateral, isosceles, and scalene).

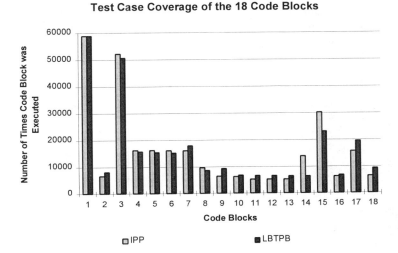

Figure 4.6 – Distribution of test cases over the 18 code blocks in trityp-easy.

FUTURE WORK

This chapter describes the use of Goldberg's SGA with the reproduction, crossover, and mutation operators. There are other operators that should prove helpful to this work. These include marriage restriction and sharing to help create species of test data.

As mentioned earlier, a C language parser is required to produce the test data generator utility. Codings appropriate for other elements of the C language need to be developed (enumerated types, arrays, pointers, etc.). Some of the authors cited earlier have begun work in this area. Inclusion of their findings in the proposed utility may prove beneficial.

The LBTPB fitness function is a special case of a more general algorithm which considers unique nodes all along the thread of execution. This general case algorithm needs to be developed further and evaluated.

One goal of testing is to minimize the number of test cases required. The work described here creates stable sub-populations of redundant test cases. To minimize tester time, a method is needed to take the set of test data produced and find the minimum number of members that provide 100% line coverage. This problem is similar in nature to the Traveling Salesman problem and should be solvable with a second genetic algorithm performing a pure minimization.

CONCLUSIONS

Genetic algorithms are useful tools for generating sets of unit test provided certain conditions are met. These include using a sufficiently large population size to allow a stable and optimal population of test cases to form at the end of the run. A population size of ten times the McCabe cyclometric complexity is recommended.

The LBTPB fitness algorithm helps the GA converge sooner and distributes test cases more evenly over the module under test than the IPP fitness algorithm. Both fitness functions tested eventually converge given an adequate population size. The LBTPB fitness algorithm accomplishes this by rewarding test cases executing unique code.

A small bonus value aids convergence. Large bonus values are disruptive. This work suggests that a bonus of 5 fitness points helps the genetic algorithm search for better test cases without causing loss of coverage due to overzealous replacement.

REFERENCES

[1] H. H. Sthamer, B. F. Jones, and D. E. Eyers (1994). Generating test data for ADA generic procedures using genetic algorithms. *Proceedings of ACEDC 1994,* 134-140, Plymouth, UK: University of Plymouth.

[2] S. Xanthakis, C. Ellis, C. Skourlas, A. LeGall, S. Katsikas, and K. Karapoulios (1992). Application of genetic algorithms to software testing. *Proceedings of the 5th International Conference on Software Engineering,* 625-636, Toulouse, France.

[3] A. L. Watkins (1995). The automatic generation of test data using genetic algorithms. In I. M. Marshall, W. B. Samson, and D. G. Edgar-Nevill (Eds.) *Proceedings of the 4th Software Quality Conference,* **2**, 300-309, Dundee, UK.

[4] D. E. Goldberg (1989). *Genetic algorithms in search, optimization, and machine learning.* Reading, MA: Addison-Wesley Publishing Company, Inc.

[5] T. McCabe (1976). A complexity measure. *IEEE Trans. on Software Engineering,* **SE –2** (4), 308-320.

CHAPTER 5

OPTIMIZATION OF A POROUS LINER FOR TRANSPIRATION COOLING USING A GENETIC ALGORITHM

Jeremy S. Lang
Department of Aerospace Engineering and Mechanics
129 Antioch Road
Somerville, AL 35670
e-mail: jlang@eng.ua.edu

ABSTRACT

In recent decades, rocket engine cooling has largely developed into a regeneratively cooled field. However, a resurgence of old concepts is playing a new role in current design ideas. The idea of transpiration cooling, mostly abandoned after development in the 1950s and 1960s, is once again being explored with new technology and materials. The technology presented here involves using a genetic algorithm to develop and optimize a liner for use in a proposed rocket engine. This liner allows an adequate amount of coolant to impinge on the hot-gas flow while maintaining a low injection velocity and pressure drop across the liner. The parameters the genetic algorithm manipulates include the height and width of the coolant channels behind the porous thrust-chamber wall, the porosity of the porous wall, and the diameter of the uniform capillaries in the wall. The genetic algorithm incorporates three main operators: roulette-wheel selection, single point crossover, and basic mutation. The genetic algorithm proves to be a successful tool in optimizing the coolant system with a proportional increase in fitness values over the generations created.

NOMENCLATURE

ϕ	porosity	C_{pc}	coefficient of specific heat for the coolant
ρ	density		
A_{open}	open area in the porous material	$f(x)$	fitness function to be maximized
$A_{surface}$	area of the entire liner surface	$g(x)$	fitness function to be minimized
A_{tc}	area to be cooled	g	gravity
c^*	characteristic velocity	hg_x	hot-gas transfer coefficient
C_{max}	maximum value of a fitness parameter	n	number of capillaries
		D	diameter of the capillaries

D_t	hydraulic diameter	Pr	Prandtl Number
L	length of the capillaries	R	nozzle radius of
Q	volumetric flow rate		curvature at throat
ΔP	change in pressure	T	temperature
	across porous surface	V	injection velocity
p	static pressure	$\dot{m}$	mass flowrate

SUBSCRIPTS

co	coolant	wg	wall-gas
aw	adiabatic wall		

INTRODUCTION

Basis for Work

Research in the field of genetic algorithms (GAs) has expanded exponentially in a relatively short period. These search algorithms have been applied in a diverse array of disciplines, and the benefits resulting from these efforts have been numerous. Thus, GAs have the potential to benefit the area of design and analysis of liquid rocket propulsion systems. In this case, the liquid rocket propulsion system is being developed by the National Aeronautics and Space Administration (NASA) George C. Marshall Space Flight Center (MSFC) Propulsion Laboratory.

It is well known in the liquid propulsion field that current coolant systems primarily make use of regenerative, film, and ablative cooling techniques. Due to the proven technology and material development for these systems, they have been used in such designs as the Space Shuttle Main Engine (SSME, which is regeneratively cooled), and the test-bed engines for the X-34 reusable launch vehicle program (which utilize a combination of film and ablative cooling techniques.) However, the National Aeronautics and Space Administration's (NASA) current goals for future engine designs are based on the ideology "reliable, reusable, inexpensive hardware." Thus, engineers at NASA's George C. Marshall Space Flight Center are looking back to the past for alternative cooling methods which may benefit their ideology.

Since the late 1960s, the transpiration-cooled system has not been rigorously examined. This is largely due in part to the lack of materials available as well as the lack of technology to develop such materials. With the technology available today, NASA has developed a renewed interest in the transpiration-cooling concept for possible future liquid propulsion systems. The result is an effort to develop an ultra low-cost assembly by adding a porous layer of material to effectively create a transpiration cooled thrust-chamber and nozzle. The ultimate goal is to improve the knowledge of and design procedures of transpiration-cooled rocket engines for future programs.

Shortcomings in Traditional Approach to Transpiration Cooling

Before discussing transpiration cooling in detail, the general definitions of each type of cooling technique mentioned must first be briefly defined.

> Regenerative cooling – A cooling method in which one or more of the propellants are fed through passages in the thrust-chamber wall to provide cooling before being injected into the combustion chamber.

> Film cooling – A cooling method in which a thin film of coolant or propellant is introduced through the injector. Additional coolant or propellant may be introduced through orifices around the thrust-chamber wall or near the throat. This method has also been used in conjunction with regenerative cooling.

> Ablative cooling – A cooling method in which a wall material on the combustion-gas-side of the thrust-chamber is deteriorated by means of melting, vaporization, and chemical changes in order to dissipate heat. In addition, the ablative material usually serves as a good thermal insulator, keeping the heat transmitted to the thrust-chamber wall to a minimum.

> Transpiration cooling – A cooling method which introduces a coolant through a porous wall at a sufficient rate to maintain the desired wall temperature for the system.

In the late 1950s through the 1960s, engineers at NASA and other organizations analyzed various coolant designs for the budding space program. One such idea was a transpiration-cooled system which developed from the flow analysis of fluids through porous media. Kelley and L'Ecuyer's paper [1] about transpiration cooling theory and application details several attempts at preparing analytical equations for the coolant required for a specified set of hot-gas and coolant parameters. For instance, Kelley and L'Ecuyer detail some early work by Rannie [2] in reference to an oversimplified model of the basic theory behind transpiration cooling. Kelley and L'Ecuyer go on to explain others who have taken up Rannie's simplified model and developed it further. A primary example is the theory of Rubesin [3] which included accounting for skin friction (and heat transfer) due to transpiration cooling. The opening segment of Kelley and L'Ecuyer's paper deals primarily with an overview of the theoretical work performed on transpiration cooling up to 1966.

Earlier in the same year, Terry and Caras [4] of the research branch of the U.S. Army Missile Command, Redstone Arsenal presented a paper on the latest theoretical and experimental work on transpiration cooling at that time. Similar to Kelley and L'Ecuyer's paper, Terry and Caras presented the current (1966) status of transpiration cooling along with the latest technology involved with the materials used for transpiration cooling. Although not directly citing individuals in the field, many of the theoretical developments given correspond to those of Kelley and L'Ecuyer's work.

Studies were also conducted experimentally using the technology available at that time. Referring to a work presented earlier, Kelley and L'Ecuyer [1] discuss the experimental work that had been performed in the area of transpiration cooling before 1966. The primary benefit of this section of Kelly and L'Ecuyer's paper was to compare empirical formulae with theoretical concepts. Similarly, Terry and Caras [4] analyzed the empirical data developed prior to 1966. However, their analysis deals primarily with material choice for the porous medium, coolant choice, and comparisons based upon these choices.

The papers presented in the previous paragraphs all lead to one main conclusion: there has been little development of transpiration cooled systems such as those described in this project since the late 1960s. Barring some minor developments in the 1980s and 1990s, little evidence has appeared as a breakthrough in the technology for transpiration cooling. Several of the papers discuss explicitly the lack of adequate, uniform materials and analytical tools necessary to make such a cooling technique effective. However, with current technologies, these hindrances no longer necessarily apply. These technologies include new uniform porous materials and simpler yet more effective designs. Improvements also exist in the state of computer design and optimization techniques which could benefit the study of transpiration-cooled systems greatly.

The Genetic Algorithm Approach

Through various conversations with representatives at MSFC's propulsion lab as well as Dr. Charles Karr, no evidence of research has been found in the area of GAs and transpiration cooling of rocket thrust-chamber assemblies. One piece of information obtained relating to transpiration cooling is found in a 1994 paper presentation by Kacynski of Lewis Research Center and Hoffman [5] of Purdue University. The primary discussion in this paper was the development of a computer code called Tethys.

Tethys is a modification of an earlier code known as Proteus. The code consisted of a, "Complete multispecies, chemically reacting and diffusing Navier-Stokes equations, including the Soret thermal diffusion and Dufour energy transfer terms." [5]. Tethys is basically an analytical tool for predicting flow characteristics, including Mach number and temperature profiles, along a chemically reacting rocket engine. This paper provides a description of the assumptions, equations, and conditions used in the development of the Tethys code. It also provides a comparison of the Tethys predictions with experimental tests of a film-cooled and transpiration-cooled rocket engine. Although providing great insight into the method with which to undertake the current project, the Tethys code relies on user-provided information and numerical computations. Unlike a GA, it does not perform search and optimization. It might, however, develop into a beneficial coding scheme for a GA.

As a result, a GA has the potential of creating an efficient, expeditious method of generating initial design dimensions for the coolant system of a transpiration-cooled liquid rocket engine. Using the hot gas flow data, as well as required coolant flowrates, the genetic algorithm is capable of varying the required design parameters in a way to create an efficient, low cost thrust chamber assembly for use in future rocket applications.

TRANSPIRATION COOLING

The term transpiration cooling can be described by the following definition which states, "Transpiration (or sweat) cooling involves low speed injection of coolant through a porous wall." [6] Transpiration cooling provides at least two principal heat reducing mechanisms. First, heat is absorbed by the cooling fluid (conduction) as it travels through the coolant channels against the direction of heat flow, from a reservoir to a surface of higher temperature and consequently lowers the wall temperature. Secondly, as the fluid emerges from the surface of the wall, it forms an insulating layer, or blanket, between the surface of the wall and the hot gas [3]. This insulating blanket is capable of reducing skin friction on the surface.

The goal of the current project is to design and analyze an engine that utilizes transpiration cooling. The basic concept here is to use a porous material as a liner through which the engine coolant can "seep" through to cool the thrust-chamber and nozzle. As can be seen in Figure 5.1, the coolant flows through channels behind the porous material. The coolant then flows with low velocity through the porous material to cool the hot gas side of the porous wall.

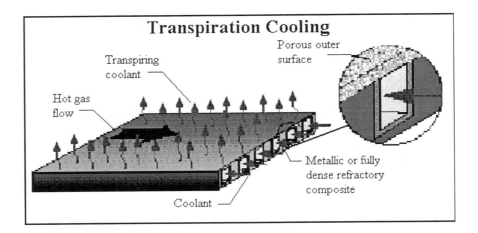

Figure 5.1 – Basic concept of transpiration cooling through a porous material.

PROJECT GOAL

The primary goal of this project is to use a GA to optimize the design of a porous liner for a liquid rocket propulsion system. Now that the concept of transpiration cooling has been defined, the details of how GAs can assist in the design of such a system can be undertaken. This discussion involves the determination of the primary optimization parameters as well as the computer model and fitness function necessary to achieve this optimization.

In order to adequately design the transpiration cooling system, several parameters need to be adjusted to provide optimum performance in three categories. The parameters include items such as the material choice for the liner, the porosity of the liner material, the hole size for the pores, and the dimensions of the coolant channels which bring the coolant to the porous liner. Each of these variables affects certain aspects of the performance of the liner and the engine. The fitness function for this GA will be based on these performance aspects.

The major performance criteria selected for the GA are the following:

1. Minimize the pressure drop across the coolant circuit.
2. Minimize the injection velocity of the coolant from the porous liner into the hot-gas flow.
3. Develop a simple, yet effective design.

The coolant in the injector is delivered through a coolant circuit behind the hot-gas wall. In order to reduce the number of design parameters considered in the problem, a copper material has been chosen for the porous liner. This material, developed by Alabama Cryogenic Engineering (ACE), can be formed with cylindrical, parallel, microscopic capillaries of varying sizes and porosities. These uniform capillaries greatly simplify the problems associated with naturally porous materials. The material simply adds the consideration of hole size and spacing. Along with the sizing of the coolant channels, these variables must be optimized to provide an adequate amount of coolant flow while preventing too large or too small a pressure drop in the coolant system. Considerations must also be made for the required low injection velocity of the coolant.

There are varying definitions for the porosity of a material. Experiments are usually required to obtain an accurate estimate of porosity for naturally porous materials. However, the uniform profile of the material presented here greatly simplifies this problem and lends itself to the following definition for porosity. The porosity of a material such as this can be described by the following equation:

$$\phi = \frac{A_{open}}{A_{surface}} = \frac{n \times D^2 \times \pi}{4 \times L^2} \tag{5.1}$$

The maximum/minimum values for the porosity are 90% and 0%, respectively. These values are set by ACE.

The important point to note here is that the porosity is the ratio of the open area of the material to its total surface area. Thus, porosity is a function of the capillary diameter and the number of capillaries. Porosity will also be important in other factors affecting the fitness function. Another important item to note is that the porosity can be changed as necessary along the length of the assembly. However, for simplicity, the changing of porosities has been limited to three distinct regions along the thrust-chamber and nozzle.

The capillary diameter is another parameter with limits set by the material manufacturer. In this case, the maximum/minimum values for the capillary diameter are 1000 micrometers and 5 micrometers, respectively. This range will be mapped into the alphabet used in the multi-parameter, fixed-point, binary coding scheme described later.

Also of importance in this design is the thickness of the liner material. An important fact to note is that as the liner increases in thickness, the weight of the entire system increases. The thickness also affects some of the important parameters listed earlier in the problem statement. Another parameter affecting the problem is the associated cost increase that occurs when the consideration of varying thickness is included. The range of values for the thickness of the material is a maximum/minimum of 1 inch and 0.125 inches, respectively.

The final two parameters which are critical in the design of the coolant system are the height and width of the coolant channels that flow behind the porous medium. These channels can be used to assist in the regulation of how much coolant flows to certain regions of the thrust-chamber assembly. The maximum/minimum dimensions for these parameters are 0.250 inches to .125 inches.

One consideration in the actual design of a transpiration-cooled liner is the pressure drop from the injector to the hot-gas wall. The pressure must be large enough to ensure that pressure fluctuations will not cause combustion to occur in the liner and coolant channels. A literature search was conducted to determine relations governing drops in pressure across a porous boundary. This resulted in several equations relating pressure drop to other fluid parameters. The equation used in this case will be Equation (5.2). This equation is developed using a model of identical parallel capillaries in the material. Using the definition of porosity (Equation 5.1), the pressure drop can be related to the geometry of the material, the flowrate of the coolant, and the fluid properties of the coolant.

$$Q = \frac{n \times D^4 \times \pi \times \Delta P}{128 \times \mu \times L} \tag{5.2}$$

Equation (5.2) contains the parameters that were described above as being important variables for the GA to consider. These parameters are the diameter of the capillaries, D, and the thickness of the capillaries, L. Thus, the GA will need to manipulate the thickness of the material as well as the diameter of the capillaries to develop an optimum solution.

The pressure drop across the porous liner is to be minimized. The pressure drop must satisfy two requirements. The first condition is based on a general engine design assumption of a maximum pressure drop of 20% of the chamber pressure. The second condition is a minimum pressure drop of 5% of the chamber pressure to allow for pressure fluctuations in the engine nozzle.

The injection velocity of the coolant is extremely important with respect to the hot-gas flow. If the coolant is injected at too high a velocity, it will create disturbances in the hot-gas flow field, causing decreased performance and increased structural stress. Thus, the injection velocity, calculated by Equation (5.3), must be low enough to avoid such a circumstance.

$$V = \frac{\dot{m}_{co}}{\rho_{co} \times A_{open}} \tag{5.3}$$

Equation (5.3) shows the inverse relationship between the injection velocity and the open area. Thus, the porosity and capillary diameter variables will be optimization factors with respect to the injection velocity.

In summary, the objective of the GA is to select values of channel height and width, porosity, and capillary diameter such that the coolant system effectively cools the thrust-chamber with a minimum pressure drop and low velocity through the porous wall.

PRELIMINARY CALCULATIONS

ODE Computer Code

A beneficial tool in the calculation of properties used in this project is a Lewis Research Center heat transfer program provided by MSFC entitled ODE [7]. This code is used to generate the hot-gas properties along the thrust-chamber nozzle assembly at desired locations. Using the engine's mixture ratio, inlet pressure, and the desired locations along the assembly, expressed as area ratios, the program provides the properties necessary for calculation of the required coolant flowrates. However, care must be taken to run the ODE program with subsonic area ratios from the chamber to the throat and supersonic area ratios from the throat to the exit.

Mass Flow Calculations

In order to obtain the coolant flowrates for the coolant system design, the hot-gas flow must first be examined. Here, formulae are given to determine the coolant flowrate required to cool the test engine to a wall-gas temperature of 1000 °F.

The given information for this procedure is the shape coordinates of the liner. Also included are the required operating conditions for chamber pressure and the desired wall gas temperature. The equation to determine the mass flow rate is as follows:

$$\dot{\omega}_{TC} = \frac{A_{TC} \times hg_x \times ((\frac{T_{aw} - T_c}{T_{wg} - T_c}) - 1)}{C_{pc}} \qquad (5.4)$$

Using the shape coordinates given by the liner design, the surface area to be cooled can be estimated using a method of conical areas. Knowing the equation for the surface area of a cone, an approximate value can be determined for the total surface area of the liner design.

The values for hg_x are given by Equation (5.5). The value for C_{pc} can be located from a table of hydrogen properties. Care must be taken to get a C_{pc} at a pressure above chamber pressure so that the hydrogen flows in one direction.

$$hg_x = \left[\frac{0.026}{D_t^{0.2}} \left(\frac{\mu^{0.2} C_p}{Pr^{0.6}} \right)_{ns} \left(\frac{(p_c)_{ns} g}{c^*} \right)^{0.8} \left(\frac{D_t}{R} \right)^{0.1} \right] \times \left(\frac{A_t}{A} \right)^{0.9} \sigma \quad (5.5)$$

where

$$\sigma = \frac{1}{\left[\frac{1}{2} \frac{T_{wg}}{(T_c)_{ns}} \left(1 + \frac{\gamma - 1}{2} M^2 \right) + \frac{1}{2} \right]^{0.68} \left[1 + \frac{\gamma - 1}{2} M^2 \right]^{0.12}} \quad (5.6)$$

The temperature requirements for the liner are both calculated and determined from desired information. The wall gas temperature desired is entered into the program. The coolant temperature for the hydrogen can be entered from design values for the coolant system. The next thing to calculate is the adiabatic wall temperature.

In order to calculate T_{aw} for the liner, the engine requirements are used in the ODE code. This code will take the mixture ratio, required fuels, and necessary area ratios, and use these values to determine various flow information at each specified point in the liner's shape. The object of this is to determine the recovery factor at each increment in the liner so that T_{aw} can be found using the following equation:

$$T_{aw} = T_{com} \times r_c \quad (5.7)$$

In this equation, r_c is the recovery factor calculated as follows:

$$r_c = \frac{1 + Pr^{\frac{1}{3}} \times (\frac{\gamma - 1}{2}) \times M^2}{1 + (\frac{\gamma - 1}{2}) \times M^2} \quad (5.8)$$

This will provide the needed values to calculate r_c. Once the r_c is determined, T_{aw} can be calculated at each increment. These calculations complete the necessary values needed to determine the mass flow rate at each increment. These mass flow rates can be added at each point, assuming no losses, and a total mass flow rate can be determined.

GA PARTICULARS

For this project, only the basic reproduction, crossover, and mutation operators are used. The program allows the user to choose the probability of crossover and mutation. The reproduction operator is based on roulette wheel selection. Additionally, the single-point crossover operator is a random choice across the entire string of the three optimization parameters.

The porosity, capillary diameter, liner thickness, and channel height and width are the optimization parameters for this system. The coding scheme for the GA manipulates these parameters with the basic genetic operators in order to generate a near optimum solution for the coolant system design.

The most important variables of the design parameters are the porosity and capillary diameter. These two variables affect almost every factor used in developing the fitness function for transpiration cooling. Because of their important bond to one another, the capillary diameter and porosity are set as the first two parameters in the chromosome string. The channel height and width are placed at the third and fourth positions in the parameter string. The less influential liner thickness will be set as the final parameter in the string.

The GA is set up to allow the user to determine the number of bits in a binary system to be manipulated for each parameter. The range for each of the variables is also a user-defined input. The number of bits chosen for each parameter determines the precision of the parameter. This format is known as a multiparameter, mapped, fixed-point coding scheme. Using a section of code similar to that of the multiparameter coding used in Goldberg's text [8], the number of bits discretizes each parameter between the minimum and maximum values for that parameter. However, a limit of ten bits is set for each parameter to avoid excessive destruction of schemata under the crossover operator.

The fitness function can be described by the two primary concerns stated earlier in the proposal. In order to optimize the system, the genetic string must satisfy the constraints set by the pressure drop and the injection velocity. The pressure drop needs to remain as close to the minimum as possible without going below the range of stability. The injection velocity will need to remain low so that the capillary diameters remain realistic.

The computer model used to evaluate the input parameters consists of the following parts:

1. The fluid properties for the combustion and coolant fluids
2. The transport properties for the combustion gases
3. The required coolant flowrate for adequate cooling of the chamber
4. The chamber pressure and wall gas temperature conditions

Parts 1 and 2 listed above are attained from the ODE code. The coolant properties, chamber pressure, and wall gas temperature are constants set by the design team. The coolant flowrate properties are attained through the methods mentioned in the mass flow calculations section. A computer model of the thrust-chamber and nozzle assembly is created for use as the fitness function in the GA.

Using the computer model data presented in the previous paragraphs, the fitness function must be developed to optimize the porous liner for the thrust chamber. The format set up for this particular GA is a maximization function based on the pressure drop and injection velocity for the liner. The details for this maximization are given below.

The pressure drop constraint is a simple minimization function, $g(x)$, with an additional upper limit constraint, C_{max}. In other words, the optimum solution must remain below the maximum allowable pressure drop while remaining as close to the minimum mark as possible. The maximum and minimum allowable pressure drops were given in the previous section. The general equation for the minimization of a function under a maximization scheme is

$$f(x) = C_{max} - g(x) \tag{5.9}$$

where C_{max} is the maximum value of the function $g(x)$, $g(x)$ is the function evaluation to be minimized, and $f(x)$ is the resulting maximization function. Using this equation,

> If $g(x)<C_{max}$, then $f(x)$ is maximized as $g(x)$ gets smaller
> If $g(x)>C_{max}$, then $f(x)$ is multiplied by a penalty function

In this case,

> $g(x)$ = pressure drop
> C_{max} = maximum pressure drop

The pressure drop must also be greater then 5%. Thus,

> If $g(x)>5\%$, then $f(x)$ is maximized as $g(x)$ gets smaller
> If $g(x)<5\%$, then $f(x)$ is multiplied by a penalty function

The injection velocity constraint is a minimization problem as well with a specified upper limit. Once again, Equation (5.9) is used where,

> $g(x)$ = injection velocity
> C_{max} = maximum injection velocity

A conditional statement is set up to ensure that the injection velocity remains below the upper limit. So,

If $g(x)<C_{max}$, then $f(x)$ is maximized as $g(x)$ gets smaller
If $g(x)>C_{max}$, then $f(x)$ is multiplied by a penalty function

Once each constraint's function is determined, the functions are added together to analyze the best total fitness values. Fitness scaling for the various constraint functions is achieved via a weighting constant for each function evaluation. The weighting values will be determined through testing. The result is the fitness equation, Equation (5.10).

$$TotalFitness = W_1 f(x) + W_2 g(x) \qquad (5.10)$$

Following this evaluation, reproduction, crossover, and mutation are carried out in the usual manner. At the beginning of the program, the user is asked to enter a minimum number of generations to reproduce. The program will terminate under two conditions. The first is that the minimum number of generations is achieved and all of the fitness constraints specified in the pressure drop and injection velocity sections are satisfied. The second termination consideration is a hard-coded maximum number of generations set at one hundred.

RESULTS

Table (5.1) below shows a simple tabulated set of general results for the GA. These results include the weighting factors for each fitness parameter and the number of bits used for each parameter to determine discretization. Note that the weighting factor for the injection velocity is much higher than that of the pressure drop. This is due to the fact that, although the pressure drop is the more important of the two parameters, the injection velocity must maintain some adequate ratio in relation to the pressure drop so that the velocity can make some contribution to the overall GA results.

First, an analysis of the fitness values of the GA leads to the development of Figures 5.2 and 5.3 below. These charts graph a typical fitness variation as a function of the two independent fitness parameters in the system. Figure 5.2 indicates the variation of average fitness values with increasing generations. Figure 5.3 indicates the change in the summation of fitness values from generation to generation. Both of these charts show an initial sharp increase in fitness values followed by a gradual leveling out of the graph as the number of generations increases.

Table 5.1 – Genetic algorithm parameter characteristics

Weighting Factors
    Pressure Drop   W=0.1
    Injection Velocity   W=3.0
Parameter Bit Lengths
    Capillary Diameter   10 bits
    Porosity   7 bits
    Thickness   9 bits
    Channel Height   8 bits
    Channel Width   8 bits
Other GA Parameters
    Generations   50
    Population Size   50

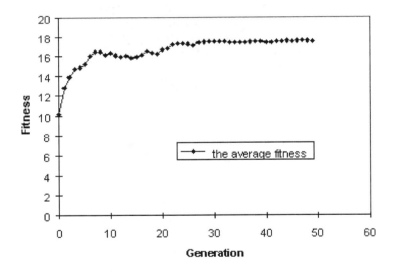

Figure 5.2 – Average Fitness Variation.

With respect to the liner thickness, one may notice a general decrease in thickness from the throat to the exit (Figure 5.4). This is primarily due to the decrease in temperature and pressure requirements toward the exit of the nozzle. Some slight errors occur near the throat possibly due to large hot-gas temperature fluctuations in this region.

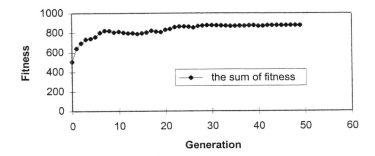

Figure 5.3 – Sum of Fitness Variation.

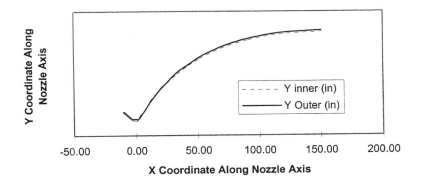

Figure 5.4 – Two-Dimensional Representation of Liner Thickness.

Analyzing Figure 5.5 below, the porosity is generally higher at the throat due to increased hot-gas temperatures in this region. At a point, shown in Figure 5.5 at about X=35 inches, the porosity drops to a value of approximately one. This severe drop is determined to be due to the large decrease in temperature of the hot-gas flow near the nozzle exit. The fluctuations in the region of the graph (Figure 5.5) near the throat are most likely due to the close relationships between the porosity and the capillary diameter.

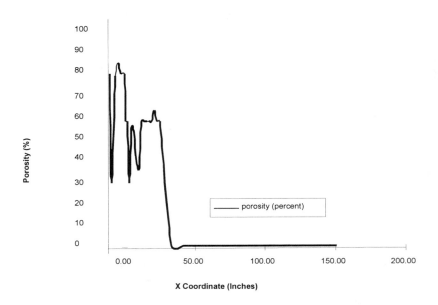

Figure 5.5 – Porosity Variation Along the Nozzle X-Axis.

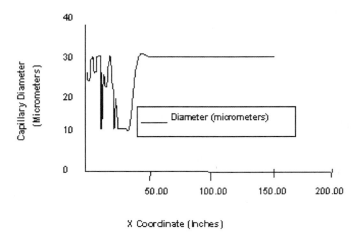

Figure 5.6 – Capillary Diameter Variation Along the Nozzle X-Axis.

The next graph, Figure 5.6, of the capillary diameter versus X-coordinate along the nozzle axis shows a similar fluctuation of values near the throat. This further solidifies the hypothesis that the porosity and capillary diameter are closely related. Looking further at Figure 5.6, as the X-coordinate increases, the capillary diameters steady out at about 31 micrometers. Once again, this can be attributed to the reduction in temperature and thus the reduction in coolant requirements towards the nozzle exit. The capillary diameter and porosity level out, leaving the optimization to occur with variations in the liner thickness.

CONCLUSIONS

Figures 5.2 and 5.3 show that the GA has achieved the goals set forth. The GA met all of the requirements and maintained a low injection velocity across the entire nozzle. Thus, success has been achieved in developing an effective liner.

Figures 5.4, 5.5, and 5.6 reveal information about the nozzle and its coolant requirements with respect to position along the axis. Generally, near the throat, the porosity and capillary diameter are closely related and most important in the optimization of that region. However, as the liner moves towards the nozzle exit, the porosity and capillary diameter level out, leaving optimization as a function of the liner thickness.

The greatest question to be considered in this effort is, "How good a solution is it?" Analytically, the solution presented here meets all of the requirements of the design team for this NASA program. The true test of the design's efficiency will lie in manufacturing and testing of a nozzle. This test liner would serve to dispel or verify the analytical solution presented here and/or the theory of transpiration cooling as developed in the 1950s and 1960s. Therefore, the next step is development of a test article to perform verification tests.

ACKNOWLEDGMENTS

First and foremost, I would like to acknowledge my wonderful wife Terisa, without whom I would not have the courage and strength to push my way through my graduate studies. Her love and support have carried me through the best and worst of times. I must also recognize my mother and father, Kenneth and Lavonda Lang, who brought me into this world and gave me the love and encouragement to pursue my dreams of joining the aerospace community. I also want to thank my sister, Nikki, for loving me even when we were mad at one another. Without these people, I wouldn't have been able to even begin such an undertaking as graduate school.

I would like to acknowledge the individuals who were so beneficial in my efforts to undertake this project. Thanks to my professors who have helped me through many quagmires along my journey through school. Thanks to Dr. L.

Michael Freeman who has been a leader, a teacher, an advisor, and a friend throughout this project. Thanks also to Dr. Charles Karr who has pushed, pulled, and dragged me through the wonderful world of genetic algorithms. Thanks also to Dr. Kevin Whitaker and Dr. Gerald Micklow for their assistance in all questions related to aerodynamics, propulsion, and fluid dynamics.

I would also like to acknowledge the assistance of NASA's George C. Marshall Space Flight Center without which I could not have worked on this project. Thanks in particular to Mr. Charles Cornelius, who has allowed me to work along with the engineers in his division of the Propulsion Lab at MSFC. Thanks also to Mr. Richard Counts who sat through many a conversation answering question after question concerning rocket propulsion. Finally, thanks to Warren Peters, an outstanding engineer at MSFC's Propulsion Lab, who helped me in almost every phase of this project.

REFERENCES

[1] Kelley, J.B. and L'Ecuyer, M.R. (1966). *Transpiration cooling - Its theory and application*: Technical Memorandum. Lafayette, IN: Purdue Research Foundation.

[2] Rannie, W. D. (1947) *A simplified theory of porous wall cooling*. Progress Report No. 4-50. Jet Propulsion Laboratory. Pasadena, California: California Institute of Technology.

[3] Rubesin, M. W. (December, 1954) *An analytical estimation of the effect of transpiration cooling on the heat transfer and skin friction characteristics of a compressible, turbulent boundary layer*. NACA Tn 3341.

[4] Caras, Gus J. and Terry, John E. (1966). *Transpiration and film cooling of liquid rocket nozzles*. Huntsville, AL: Redstone Research and Development Directorate.

[5] Hoffman, Joe D. and Kacynski, Kenneth J. (1994). *The prediction of nozzle performance and heat transfer in hydrogen/oxygen rocket engines with transpiration cooling, film cooling, and high area ratios*. National Aeronautics and Space Administration.

[6] Hendricks, Dr. John B. (1996). *An innovative concept for cooling liquid propellant rocket combustion chambers*: Final Technical Report. Huntsville, AL: Alabama Cryogenic Engineering, Inc.

[7] Lewis Research Center. *ODE code*. Public Domain Software.

[8] Goldberg, David E (1989). *Genetic algorithms in search, optimization, and machine learning*. Reading, MA: Addison-Wesley Publishing.

CHAPTER 6

GENETIC ALGORITHMS FOR CONSTRAINED SERVICE PROVISIONING

Phil Benjamin
Data Services Integration Division
AT&T
101 Crawfords Corner Road
Holmdel, NJ, 07733
e-mail: pmbenjamin@att.com

ABSTRACT

The focus of this chapter is an industrial decision problem involving the provisioning of services subject to a variety of constraints. Provisioning involves the allocation of generally scarce and costly resources to services subject to a variety of constraints, including both resource constraints and inter-service constraints. In addition to technical constraints on resource utilization and cost minimization, there are also often fuzzy and even contradictory criteria about the "quality" of a proposed provisioning. The approach taken here allows a dualism between satisfying soft constraints and meeting service objectives. Objectives include minimizing cost, meeting individual service objectives, and providing both service redundancy and isolation. A practical approach to this problem requires finding a "good enough" solution to a provisioning problem in a "reasonable" amount of time. A related problem is to reallocate resources when a new service is added to the collection of currently offered services. A genetic algorithm has been implemented to solve an abstraction of the problem of constrained resource provisioning. The implementation is quite general and can be used to solve a wide variety of constrained provisioning problems. This chapter discusses the approach, the implementation, and the evaluation of the approach using some sample data.

PROBLEM FORMULATION

Problem Background

It is desired to offer network-based computer-telephony services to customers. A "service" in this context is generally a computer resident complex (programs, databases, links to customer-based resources, etc.) which is accessible via a telephony network and is used to provide an economic service to a customer. The services are "provisioned" by assigning them to resident computer and telephony resources. The "provisioning problem" is to make a set of interrelated decisions about how to allocate resources to services.

There are both technical and business constraints to consider in the provisioning process. On the one hand, a certain contractual commitment may be in effect which requires a certain level of service to exist. There is little latitude in meeting this constraint. On the other hand, a service may be provisioned at a level to maximize net revenue. This decision often involves "soft" market data and guesses about revenue potential, especially for new services.

In general, services are not provisioned independently from each other. Resource sharing may be feasible for some sets of services but others must be provisioned so that they do not share resources. This may be for contractual reasons (e.g., guaranteed service levels) or for risk minimization (i.e., "not putting all eggs in one basket").

Providing a multitude of services must be done, in the short term, using existing resources. The available resource levels can be "stretched" but usually with some penalty such as excessive cost. So resources, like services, are subject to both "hard" and "soft" constraints.

Provisioning Procedures in Practice

There have been attempts to automate the provisioning of telephony services. A common problem encountered is the accumulation and access to appropriate data needed to make decisions. Much of the data needed to make provisioning decisions is surprisingly "soft". Estimated traffic volume and patterns may be off by a factor of two or more. Usage of telephony resources should be reliable so critical services are often over-provisioned in order to (try to) guarantee a certain level of service even under adverse conditions.

In practice, a great deal of experience and insight is necessary in order to make good provisioning decisions. Automated decision procedures have not had a good track record and are not widely used. Many projects use resident "provisioning experts" who have sufficient experience with past provisioning practices and "rules of thumb" that they can often make reasonable, if not optimal, provisioning decisions. A typical scenario is that a person makes an informed decision about how to provision a new service and then considerable monitoring is done after the service "goes live", to see whether the human decision is working well. Frequently, adjustments will be made after the service is deployed. This process is often expensive and error-prone. Better provisioning approaches are needed.

The current effort does not attempt to model a real provisioning system due to both the magnitude and complexity of the problem. Rather, some basic ideas from genetic algorithms, as presented in Goldberg [1], are applied to the provisioning problem in order to gain an understanding of their applicability to various aspects of the provisioning process.

Problem Abstraction

In order to simplify the problem, the following abstraction is used. Let there be given entities of the following types:

- Resources: Res = { R(1), R(2), ... }
- Services: Svc = { S(1), S(2), ... }

A "provisioning" is specified by a function P(r,s). P(r,s) is a real-valued function with domain the Cartesian product: Res x Svc. P(r,s) measures the amount of service "s" to be provided using resource "r".

In graph theory terms, the problem can be represented using a bi-partite graph:

$$Res \Leftrightarrow Svc$$

where there are edges between resources and services. The "provisioning" is an "edge-labeling" using real-valued labels.

An interesting aspect of this problem is the duality between constraints and objectives. In one view, the problem is to minimize the "cost" (resources) needed to provide a given level of service. This view regards the services as providing fixed constraints with the objective function being a measure of the resources used to provide that service.

A complementary viewpoint is to maximize the "service" (strongly correlated with profit hopefully!) which can be provided using a fixed set of resources (cost). In this view, the resources are the fixed constraints and the objective function is a measure of the level of service provided using those resources.

The abstraction investigated here has some interesting properties:

1. Traditional approaches to this problem (e.g., using linear programming) must commit to one view or the other up front. (i.e., "fixed cost with maximum service" or "fixed service with minimum cost"). This "premature commitment" decreases the utility of the computation. The approach here, using a genetic algorithm, allows complete flexibility by weighing the penalty for NOT achieving each "requirement" separately. As an example, suppose three services, S1, S2, S3 are to be provisioned using two resources R1, R2. Parameters of the genetic algorithm can be adjusted to give S1 and R2 (say) quadratic penalty functions and the other services and resources only linear penalty functions. The effect of this is to say "the solution MUST meet service constraints for S1 and resource constraints for R2 and it's DESIRABLE to meet the other service and resource constraints."

2. Measurement of resource utilization as well as the level of service provided are often "soft." For example, level of service might include some "customer satisfaction" data gained from a survey. Traditional approaches (e.g., linear programming) do not handle "soft" data well. Our genetic algorithms approach deals (roughly) equally well with underconstrained, tightly constrained, or overconstrained problems. As discussed in the Investigations section below, the genetic algorithm did not always find the optimal solution but did always find a "reasonable" solution. In contrast, linear programming approaches (when all constraints are linear) do well to find feasible and optimal solutions when they exist, but, in general cannot handle overconstrained problems (which often occur in the "real world").

3. For a given provisioning, there may be several ways to measure the cost and the service provided. As mentioned above, "service isolation" is a worthwhile goal. "Resource sharing" is another worthwhile goal but generally diminishes service isolation. It is useful to have a model which can take multiple objectives into account. A similar statement can be made about the measurement of resource costs. This is implemented in a business setting by having a proposal reviewed by multiple people or departments. The final plan is generally a compromise between various views. A variation of the basic genetic algorithm was implemented to model this business decision process. Both a "primary" and "alternate" set of constraints were presented. The genetic algorithm was designed to find solutions that are "reasonable" solutions to each set of constraints simultaneously.

Summary

A genetic algorithm approach to an abstraction of the "Provisioning Problem" as described above has been found to be useful. A review of some related work is given in the next section. The actual genetic algorithm formulation of the problem is discussed in the third section. Some detail on the computer implementation is then provided, and results of four different types of problems investigated are provided.

RELATED WORK

The optimization problem considered here has some aspects in common with other problems which have been studied:

-   A potentially large number of constraints. For the simplified problem discussed here, the complexity of the constraints will be minimized but in the "real world" there are diverse constraints.
-   Multiple objectives for the optimization.

The Web-based Genetic Algorithms Digest [2] is helpful as a forum for an informal exchange of ideas and was consulted for this project.

Michalewicz [3] is an excellent book that describes many genetic algorithm issues and problems. Chapter 9 suggests a representation for the transportation problem (quite similar to the problem of this project) which uses a special initialization algorithm (p.145) and a permutation technique which works quite well for the current problem.

Cox et al. [4] consider the problem of designing dynamic routing in communications networks. This work is applicable to the current project in that similar constraint (meeting service needs without using excessive resources) considerations exist. Also, since the problem is to find a *dynamic* solution, it starts from a given routing and looks for improvements as call traffic fluctuates.

Davis and Coombs [5] apply genetic algorithms with some interesting approaches to constraints. They developed Stochastic constraints (things are "OK" if the constraint is satisfied "most of the time," defined in the paper), Ice Age constraints (constraints are only applied during "Ice Ages," i.e., rarely, instead of every generation), and "LaMarck Operators" which handle constraints by fixing them up occasionally. They also applied Object-Oriented programming techniques to genetic algorithms for handling constraints. Their approach to "Ice Age" constraints was used in one of the investigations of this project.

Surry and Radcliffe [6] treat multiple constraints as a Pareto optimization problem, with satisfaction of each constraint taken as another objective function. A special method is developed: COMOGA (Constrained Optimization by Multi-Objective Genetic Algorithms). This entails the calculation of constraint violations, Pareto ranking, fitness evaluation, and selection. The selection operation "compromises" between picking the "best" members (based on fitness) and the "best" members based on constraint satisfaction. The approach for this investigation uses a variation of the Pareto approach for tournament ranking as discussed in the Investigations section below.

Michalewicz [7] is a survey article that outlines many of the popular methods of dealing with constraints. One interesting method, useful in dealing with *convex* problems, is to avoid constraint violation by considering crossover as a linear combination of (feasible) individuals.

GENETIC ALGORITHM MODEL

Genetic Algorithm Approach for this Problem

The strategy employed here is based on a genetic algorithm approach as described in Goldberg [1] with certain variations. The remainder of this chapter assumes familiarity with the concepts and terminology of genetic algorithms as presented in Goldberg [1]. This section describes the formulation of the chromosomes, the methods of selection, reproduction, mutation, and the fitness functions.

Problem Size

For this approach to be useful commercially, it should be successful on problems using up to 100 resources and 100 services. Here, only small to medium size problems are considered, with up to 10 resources and 10 services.

Chromosome Formulation

Chromosomes are a two-dimensional matrix P (for "provisioning"). The rows of P are the resource pools to be used. The columns of P are the services to be provisioned. P(r,s) represents the amount of service "s" provided through resource "r". The length of the chromosome is |R| x |S|. Thus, a typical chromosome might be:

(5 2 0 0 0 3 7 8 1 9 4 2 2 3 0 1 6 8 4 4 3)

representing the following provisioning:

Resource	Services						
	Svc #1	Svc #2	Svc #3	Svc #4	Svc #5	Svc #6	Svc #7
Res #1	5	2	0	0	0	3	7
Res #2	8	1	9	4	2	2	3
Res #3	0	1	6	8	4	4	3

Allele Formulation

The entries in the chromosomes are taken here to be small integers. As discussed below, alleles will be preserved under crossover.

The approach taken here is a compromise between the traditional genetic algorithm approach and a linear programming approach. Traditionally, a genetic algorithm would encode data into a binary (or small cardinality) alphabet string using discretization of the solution space. This makes the search space more manageable but also requires early commitment to a specific set of possible solutions which is overly restrictive in the current problem domain.

In contrast, a linear programming approach would generally use floating point operations to find a solution (if one exists, if all constraints are linear, etc.) to a high level of precision. This level of precision is unwarranted here, both due to the "softness" of the constraints mentioned above and due to the fact that a provisioning of services is ultimately carried out by people making commitments of resources to provide services. For example, six channels might be committed to service A, but not 6.037429 channels, even though the latter solution might be more optimal in a mathematical sense.

Mutation

Mutation of an allele will take place using a "centered" mutation algorithm. Treating each allele as a non-negative integer, a mutation will increment or decrement that integer by a small amount, chosen uniformly over an interval. The length of this interval and the probability of mutation are parameters that can be adjusted. Mutations will always produce non-negative integer values alleles. That is, a mutation which produces a negative value will be adjusted to be "0". Mutations are not explicitly constrained from growing "too large" but chromosomes with high values will generally have high penalties associated with them. Thus, the problem space is effectively limited to a compact domain.

Crossover

For this problem, crossover is done in a slightly "non-standard" manner. This is dictated by the fact that a chromosome represents a two-dimensional structure, and the crossover method should reflect that structure. The approach here is "quadrantal" (See Figure 6.1).

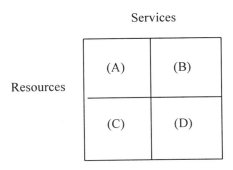

Figure 6.1 – Quadrantal crossover method.

Crossover is implemented by choosing a crossing point in the resource list and a crossing point in the services list. The two parent chromosomes are then "crossed," creating two children chromosomes by using the quadrants from each parent. The method used here is, for each quadrant, one child, chosen at random, receives one parent's version, the other child receives the other parent's version. This decision can be visualized by considering the following example table:

Quadrant:	Parent Selection	
	Child #1	**Child #2**
A	Parent #1	Parent #2
B	Parent #2	Parent #1
C	Parent #2	Parent #1
D	Parent #1	Parent #2

In this example, Child #1 has quadrants A,B,C,D from Parents #1, #2, #2, #1, respectively. Child #2 has quadrants A,B,C,D from Parents #2, #1, #1, #2, respectively. For each crossover operation, the assignment of parent quadrants to children is determined randomly. This approach has the advantage that all genetic material from both parents is preserved in some offspring.

Discussion of Crossover Strategy

Several methods of crossover were considered in this study before the current approach was adopted. One method, creating a **single** child offspring by taking the best combination of quadrants from each parent, was successful for small problems but tended toward premature convergence with valuable genetic material lost. Another approach with a fixed allocation of parents' quadrants to offspring preserved genetic material but tended to "lock" large areas together, which then led to insufficient mixing of the population. The current strategy seems to combine a good "mixing" approach with no loss of genetic material.

Fitness Parameters and Fitness Functions

A fitness value must be given to a chromosome in order to implement the selection process and, finally, to determine whether the genetic algorithm has arrived at a good solution. Generally, the objective is to minimize resource usage while maximizing service availability. This is modeled here by assigning the following parameters:

1. **Unit cost** parameter for each allele locus that measures the unit cost of providing service (s) using resource (r).

2. **Unit benefit** parameter for each allele locus that measures the unit benefit of providing service (s) using resource (r).
3. **Resource limits** are parameters that reflect the limited availability for each resource. Fitness will incur a penalty when the weighted row sum of the chromosome values exceeds the resource limit.
4. **Benefit limits** are parameters that reflect the **minimum** limits of service that should be provided. Fitness will incur a penalty when the weighted column sum of the chromosome values is less than the associated benefit limit.

The chromosome fitness value is the sum of the row and column penalties. Two further parameters are available for row and column fitness evaluation:

1. **Fitness multipliers** are used for both rows and columns. These multipliers are needed because the measurement of resource cost and service benefit may be in different units or have different priorities. Use of multipliers allows a weighted sum of these two penalties to be computed.

2. **Fitness exponents** are used for both rows and columns. Each row or column penalty may be raised to an exponent (generally 1 or 2) before being combined with other penalties. The introduction of a non-linearity in the penalty function provides a mechanism for enforcing a hard limit (minimum or maximum). For the investigations here, using exponent "2" is the "hard" limit, and using exponent "1" is the "soft" limit. The type of limit for each resource and service level can be set independently.

Selection

Since the approach here uses (positive) penalty functions, the selection criteria uses minimization. Thus, a tournament selection method is appropriate and is used here. The number of participants in each tournament and the number of winners in each tournament are problem instance parameters. For example, tournaments could be set up with five "contestants" selected at random from the population. Then each tournament could designate as "winners" the two contestants with lowest fitness (first place and second place). The numbers "five" and "two" in the example above can be varied in the implementation.

An interesting side effect of this strategy is that the absolute size of the fitness (penalty) of each chromosome is not relevant, only its relative value compared to other chromosomes. This approach may more accurately reflect the uncertainty of the fitness data in real problems.

A variation of this is used in the "multi-objective" experiment #4 described below.

COMPUTER IMPLEMENTATION

The computer code for this investigation is implemented in the "C" language and consists in its current version of about 4,000 lines of code. All experiments were run on a SUN platform running UNIX. All parameters for a run are contained in a parameters file described below.

The initial population is generated by a separate program, **randArray**, which simply generates a population of arrays of integers within certain bounds. For experiments here, the bounds 0...10 were used.

The principal program operates by reading the initial population from standard input performing the genetic algorithm according the settings in the parameters file (taken as an argument), and writing the final population to the standard output. This design is a standard UNIX approach that permits "chaining" of different variants of the algorithm into an extended algorithm and provides a convenient tool for investigation of time-varying parameters for the program as well.

For each generation, the following steps are performed:

1. **Fitness** computation
2. **Tournament** selection
3. **Mutation**
4. **Crossover**

After each step above, the complete population is scanned to see whether any individual has fitness less than the stopping fitness parameter. If so, the genetic algorithm terminates. If not, it continues up to a maximum (parameterized) number of generations.

The Parameters File

The parameters file guides the operation of the genetic algorithm. It contains all problem-specific information as well as all "tuning" parameters. Each of the entries in the parameters file is described here.

Population parameters:

- **nrow** - number of rows (resources)
- **ncol** - number of columns (services)
- **npop** - population size

Random number seed

- **seed** - random number seed. "0" is used to request a "current time" based seed.

Stopping parameters

- **stopgen** - maximum number of generations to run before stopping the algorithm
- **stopfit** - maximum value of "acceptable" fitness. Ideally, this would be "0" but for tightly constrained or overconstrained problems, setting this to a "reasonable" value aids in problem convergence.

Mutation parameters

- **muprob** - represents the probability of individual allele mutation.
- **murng** - represents the maximum range of mutation value possible. For example, by setting **murng=10**, a mutation will select a "delta" value uniformly in the range of $-10...+10$ and apply it to the allele.
- **mugenlarge** - duration (in generations) between "epochal" mutation. Every "**muperiodlarge**" generations a different set of mutation parameters is put into effect for a single generation.
- **muproblarge** - the probability of individual allele mutation during the "epochal" generation. Generally this is larger than **muprob.**
- **murnglarge** - the maximum range of mutation value during the "epochal" generation.

Tournament parameters

- **toursize** - number of participants in each tournament.
- **tourwin** - number of winners chosen in each tournament.

Fitness Parameters

All fitness parameters here have a default value of "1" unless otherwise specified.
- **rowsum** - vector of numbers representing **MAXIMUM** weighted row sums which can occur without penalty.
- **rowfact** - vector of numbers representing multipliers for row penalty values.
- **rowexp** - vector of numbers representing exponents applied to row penalty values.
- **colsum** - vector of numbers representing **MINIMUM** weighted column sums which can occur without penalty.
- **colfact** - vector of numbers representing multipliers for column penalty values.

- **colexp** - vector of numbers representing exponents applied to column penalty values.
- **cost** - array of numbers representing unit cost per allele in the chromosome. These are used as multipliers when computing row (resource) penalties.
- **bene** - array of numbers representing unit benefits per allele in the chromosome. These are used as multipliers when computing column (service) penalties.

INVESTIGATIONS

The problem examples below are all based on the genetic algorithm formulation above. All examples can be run using the same code formulation with varying values for the problem parameters.

Problem #1: Tight Constraints

The parameters file for this problem is shown here:

Population			Fitness					
nrow	5		rowsum	25	25	25	25	25
ncol	5		rowfact	1	1	1	1	1
npop	100		rowexp	1	1	1	1	1
Random Seed			colsum	25	25	25	25	25
seed	0		colfact	1	1	1	1	1
			colexp	1	1	1	1	1
Stopping								
stopgen	100							
stopfit	0							
Mutation								
muprob	0.01							
murng	5							
Tournament								
dotour	1							
toursize	5							
tourwin	1							
Crossover								
xprob	0.2							

The best (minimal) fitness for each generation is shown next:

Tight Constraints

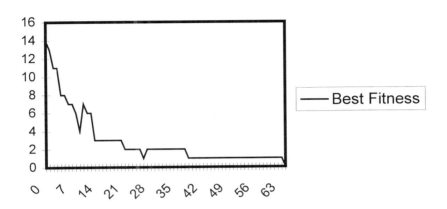

Discussion for Problem #1

This problem is small (five resources, five services) and simple (all resources and services are interchangeable). The interesting part is that the sum of the resource constraints is equal to the sum of the service constraints. Since the resource constraints are **maxima** and the service constraints are **minima**, equality of their sum means that, for a zero penalty solution, all constraints must be met exactly. Since the number of constraints (10: one for each row, one for each column) is far less than the number of variables (25 alleles per chromosome), this problem is, in general, solvable. However, the solution space can be thought of as a "thin" (15-dimensional) hyperplane in 25-dimensional space. This is actually a rather small "target" for the genetic algorithm to hit!

The genetic algorithm was able to find a zero penalty solution in 64 generations. The solution found is shown here:

	Svc #1	Svc #2	Svc #3	Svc #4	Svc #5
Res #1	2	7	7	1	8
Res #2	6	2	8	8	1
Res #3	3	6	2	5	9
Res #4	10	2	8	3	2
Res #5	4	8	0	8	5

This problem turns out to be "easy" for the genetic algorithm. It was performed as a feasibility test for the general approach described in this chapter.

Problem #2: Unequal Resources and Services

The parameters file for this problem are shown here:

Parameters File

Population		Fitness						
nrow	6	rowsum	30	30	30	30	30	30
ncol	6	rowfact	1	1	1	1	1	1
npop	100	rowexp	1	1	1	1	1	1
Random Seed		colsum	60	60	60	60	60	60
seed	0	colfact	1	1	1	1	1	1
		colexp	1	1	1	1	1	1
Stopping								
stopgen	2000	Cost						
stopfit	0		1	1	1	2	2	2
			1	1	1	2	2	2
Mutation			1	1	1	2	2	2
muprob	0.01		2	2	2	1	1	1
murng	5		2	2	2	1	1	1
mugenlarge	10		2	2	2	1	1	1
muproblarge	0.2							
murnglarge	2	Benefit						
			1	1	1	1	1	1
Tournament			2	2	2	2	2	2
dotour	1		3	3	3	3	3	3
toursize	3		3	3	3	3	3	3
tourwin	1		2	2	2	2	2	2
			1	1	1	1	1	1
Crossover								
xprob	0.2							

The best (minimal) fitness for each generation is shown below:

Unequal Costs, Benefits

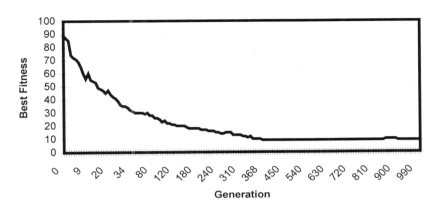

Discussion for Problem #2

The problem size is almost the same as for Problem #1 (six resources and six services) but, in this problem, it is much more difficult to "see" a solution by inspection. The genetic algorithm quickly comes near to a solution after about 300 generations (penalty 10) and thereafter, does not find further improvement.

This problem was "rigged" so that both the cost and benefits matrices (shown above) have unequal entries. Examining the **cost** array for resources, we see that resources #1 ... #3 are more efficient for providing services #1 ... #3. Similarly, resources #4 ... #6 are more efficient for providing services #4 ... #6.

The **bene** (benefits) array is also biased. Services #3 and #4 provide the most benefit per unit of resource with other services diminishing in value per resource used. Putting this together, it makes most sense (to a person!) to provide most of services #1 ... #3, with resource #3 and most of services #4 ... #6 with resource #4.

The solution found by the genetic algorithm is shown here:

	Svc #1	Svc #2	Svc #3	Svc #4	Svc #5	Svc #6
Res #1	7	6	7	4	0	0
Res #2	6	11	7	3	0	0
Res #3	9	10	13	0	0	0
Res #4	0	0	0	13	15	10
Res #5	6	0	0	2	6	10
Res #6	2	2	0	7	5	10

The genetic algorithm has approximately duplicated the "experienced provisioners" solution. This solution is not "ideal" (zero penalty) but there is no obvious (to this author!) way to improve the solution.

Problem #3: Alternate Values

This experiment is designed to investigate the "real world" problem of simultaneously meeting multiple objectives. Consider the utilization of resources as an objective autonomous system with two observers. Each observer independently places weights on the utilization of resources and the values of services provided. The objective of the genetic algorithm is to find a "Provisioning" which satisfies both observers. The experiment is as in Problem #2 above, with two separate sets of weights leading to two separate fitness values. In the parameters file below, these sets are labeled "Primary" and "Alternate."

The parameters file for this problem is shown on the following page:

Parameters File

Population	
nrow	5
ncol	5
npop	100

Random Seed	
seed	0

Stopping	
stopgen	1000
stopfit	0
stopfitalt	0

Mutation	
muprob	0.05
murng	5
mugenlarge	20
muproblarge	0.2
murnglarge	2

Tournament	
dotour	1
toursize	3
tourwin	1

Crossover	
xprob	0.2

Primary Fitness

rowsum	76	76	76	76	76
rowfact	1	1	1	1	1
rowexp	1	1	1	1	1
colsum	74	74	74	74	74
colfact	1	1	1	1	1
colexp	1	1	1	1	1

Cost

1	2	3	4	5
1	2	3	4	5
1	2	3	4	5
1	2	3	4	5
1	2	3	4	5
1	2	3	4	5

Benefit

1	1	1	1	1
2	2	2	2	2
3	3	3	3	3
4	4	4	4	4
5	5	5	5	5

Alternate Fitness

rowsumalt	76	76	76	76	76
rowfactalt	1	1	1	1	1
rowexpalt	1	1	1	1	1
colsumalt	74	74	74	74	74
colfactalt	1	1	1	1	1
colexpalt	1	1	1	1	1

Cost(alt)

5	4	3	2	1
5	4	3	2	1
5	4	3	2	1
5	4	3	2	1
5	4	3	2	1
5	4	3	2	1

Benefit(alt)

5	5	5	5	5
4	4	4	4	4
3	3	3	3	3
2	2	2	2	2
1	1	1	1	1

The following chart shows the best fitness for each generation.

Primary/Alternate (Partial)

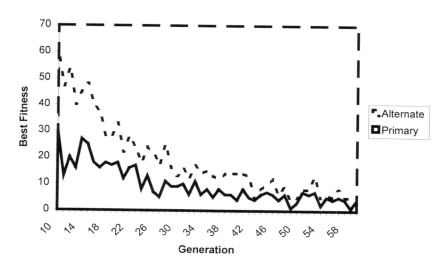

This sample shows that the algorithm is "trying" to keep both fitnesses minimized simultaneously.

In order to investigate this problem, an alternate implementation was developed. It was based on the original implementation but with fitness parameters and values existing in both primary and alternate form. This required a special stopping criteria and a special decision process to be performed for tournament selection.

The stopping criteria were simply extended: the genetic algorithm is permitted to stop only when **both** fitness criteria are met (primary and alternate), or when the maximum number of generations has been run.

The tournament selection method was more problematic. Several approaches were investigated here. The one finally adopted is as follows. Let the fitness penalties using the different weightings be **P** and **Q**. Then the combined fitness penalty is:

$$P + Q + Max(P,Q)$$

The concept behind this formulation of "total" penalty is designed so that both of the "observers" will not be overly "unhappy." To see how this works, consider two example scenarios:

1. Scenario #1: P = 2, Q = 10. In addition to P being "happy" with the fitness and Q being "unhappy" with the fitness, in "real life," Q would probably also be "unhappy" about the "unfairness" of the solution. (Q: Why should P be happier than I?) So the total penalty of 2 + 10 + 10 = 22 is assessed.

2. Scenario #2: P = 5, Q = 8. Here P is less "happy" and Q is more "happy" but, since the unhappiness is "shared," there is less total unhappiness. So the total penalty of 5 + 8 + 8 = 21 is assessed.

Irrespective of the "pop psychology" explanation, using the method outlined here leads to a solution that is able to minimize **both** fitness penalties simultaneously.

The solution found by the genetic algorithm is shown here:

	Svc #1	Svc #2	Svc #3	Svc #4	Svc #5
Res #1	3	6	6	8	2
Res #2	6	6	3	2	8
Res #3	7	3	6	1	8
Res #4	7	1	5	9	3
Res #5	2	9	5	5	4

Problem #4: Service Isolation

As discussed above, there are a wide variety of issues which must be considered by the service provisioner. Maximizing service benefits and minimizing resource costs are generally important. However, other issues are also important. Here we consider the problem of service isolation.

In order to model this problem using a genetic algorithm, a modification to the fitness function was made. In addition to the row and column penalties assessed as described above, an additional section was added to the parameters file as the following shows.

Parameters File

Population			Fitness									
nrow	10		rowsum	55	55	55	55	55	55	55	55	55	55
ncol	10		rowfact	10	10	10	10	10	10	10	10	10	10
npop	100		rowexp	1	1	1	1	1	1	1	1	1	1

Random Seed			colsum	50	50	50	40	40	40	40	40	40	40
seed	0		colfact	50	50	50	10	10	10	10	10	10	10
			colexp	1	1	1	1	1	1	1	1	1	1

Stopping

		Isolation Factor									
stopgen	4000		1	1	1	1	1	1	1	1	1	1
stopfit	0		1	1	1	1	1	1	1	1	1	1
stopfitalt	0		1	1	1	1	1	1	1	1	1	1

Mutation

		0	0	0	0	0	0	0	0	0	0
muprob	0.05	0	0	0	0	0	0	0	0	0	0
murng	5	0	0	0	0	0	0	0	0	0	0
mugenlarge	20	0	0	0	0	0	0	0	0	0	0
muproblarge	0.2	0	0	0	0	0	0	0	0	0	0
murnglarge	2	0	0	0	0	0	0	0	0	0	0
		0	0	0	0	0	0	0	0	0	0

Tournament

dotour	1
toursize	3
tourwin	1

Crossover

xprob	0.2

Here, an array, **isofact**, captures the inter-service isolation factors desired. These factors act as multipliers and are applied whenever two services share a common resource. For the example shown, all factors are "1" for Services #1 .. #3 and "0" otherwise. This causes a penalty to be assessed whenever any of these services share resources with any other services.

This problem is studied here both because it is an important "real world" constraint, and because it adds an interesting epistasis to the problem domain. In addition to "competing" for resources, certain (high priority) services are forced to be separated from each other.

This example took 1,288 generations to arrive a zero penalty solution, shown here.

	Sv #1	Sv #2	Sv #3	Sv #4	Sv #5	Sv #6	Sv #7	Sv #8	Sv #9	Sv #10
Res #1	0	0	0	1	4	19	8	3	14	2
Res #2	0	0	0	22	0	0	5	6	2	9
Res #3	0	0	0	9	14	7	2	7	10	1
Res #4	0	0	0	0	15	0	3	23	6	6
Res #5	0	0	0	8	5	9	15	1	8	0
Res #6	0	0	0	13	9	5	3	0	7	7
Res #7	50	0	0	0	0	0	0	0	0	0
Res #8	0	0	0	13	3	0	8	2	2	15
Res #9	0	0	50	0	0	0	0	0	0	0
Res #10	0	54	0	0	0	0	0	0	0	0

The solution found by the genetic algorithm shows that the isolation strategy has been achieved: Service #1 uses only Resource #7, Service #2 uses only Resource #10, and Service #3 uses only Resource #9.

CONCLUSION

A genetic algorithm approach to an industrial decision problem, resource provisioning, has been considered. A flexibly parametrized program has been developed to investigate the problem for small-to-medium size examples. The genetic algorithm approach is shown to be feasible and matches human intuition for small examples. The genetic algorithm always found a reasonable solution in a reasonable number of generations. It did not always find an optimal solution in problems where an optimal solution was known to exist.

The algorithm was applied in two basic cases. Two different extensions of the algorithm were also studied. One extension showed that simultaneously considering multiple cost and benefit factors is feasible within the same problem approach. A separate extension showed that adding a highly non-linear component to the fitness (penalty) function could also be handled easily within the same framework.

REFERENCES

[1] Goldberg, D. (1989). *Genetic algorithms in search, optimization and machine learning.* Reading, MA: Addison-Wesley.

[2] *Genetic Algorithms Digest:* http://www.aic.nrl.navy.mil/galist.

[3] Michalewicz, Z. (1992). *Genetic algorithms + data structures = evolution Programs.* New York: Springer-Verlag.

[4] Cox, L.A., Davis, L., Qiu, Y. (1991). Dynamic anticipatory routing in circuit-switched telecommunications networks. In Davis, L. (Ed.), *Handbook of genetic algorithms.*

[5] Davis, L., Coombs, S. (1987). Genetic algorithms and communication link speech design: Theoretical considerations, constraints and operators. In *Genetic Algorithms and Their Applications, Proceedings of the Second International Conference on Genetic Algorithms*, Cambridge, MA: MIT Press.

[6] Surry, P., Radcliffe, N. (1995). A Multi-objective approach to constrained optimization of gas supply networks: The COMOGA method. In Fogarty, T. (Ed.), *AISB-EC-95.*

[7] Michalewicz, Z. (1995). A survey of constraint handling techniques in evolutionary computation methods. In McDonnell, J., Reynolds, R., Fogel, D. (Ed.) *Evolutionary Programming IV*, Cambridge, MA: MIT Press.

CHAPTER 7

USING A GENETIC ALGORITHM TO DETERMINE THE OPTIMUM TWO-IMPULSE TRANSFER BETWEEN CO-PLANAR, ELLIPTICAL ORBITS

Angela K. Reichert
Department of Aerospace Engineering Sciences
University of Colorado
Campus Box 431
Boulder, CO 80303-0431
e-mail: reichera@orbit.colorado.edu

ABSTRACT

Many years ago, when the idea of space exploration was beginning to become a reality for mankind, scientists working in the field of orbital mechanics began studying the motions and maneuvering requirements for an orbiting spacecraft. Because fuel is an expensive commodity, the minimization of the amount required to maneuver a spacecraft from one orbit to another demanded high priority. The amount of fuel needed to perform an orbit transfer is directly related to the spacecraft's velocity change, or ΔV, as it maneuvers. Therefore, in order to minimize fuel usage, the ΔV must be minimized. Because this problem of determining the minimum ΔV cannot be easily solved in closed form, simplifications were usually made concerning the initial and final orbit properties. With sufficient simplifications, it became possible to solve the governing equations in order to obtain a solution.

In this chapter, a different approach to the minimum ΔV problem is proposed for the case of coplanar, elliptical orbits. Instead of trying to solve the governing equations directly, a genetic algorithm (GA) is utilized to find the optimum transfer trajectory required for minimum ΔV. With the geometry of the initial and final elliptical orbits given, the GA is employed as a "brute force" method of solving the problem, by first identifying a population of pseudo-random transfer orbits, specified by the three parameters required to define an ellipse in a given plane: the eccentricity, the semi-latus rectum, and the orientation angle of the ellipse from a given reference point. For each possible transfer orbit, the minimum total ΔV is calculated, provided that a total of two ΔV impulses are applied – the first to change the orbit from the initial to the transfer orbit, and the second to change from the transfer to the final orbit. In this chapter, the deciding factor in determining the optimum transfer orbit is the value of the total ΔV required to maneuver from the initial to the final orbit. Therefore, the equations for computing ΔV define the fitness function (the function to be optimized) by the GA.

0-8493-9801-0/98/$0.00+$.50
© 1999 by CRC Press LLC

In order to prove the validity of the solution found using a GA for a general case, the transfer orbits for three known test cases are calculated using the GA search method. These cases include (1) the transfer between coplanar circular orbits, or Hohmann transfer, (2) the transfer between coplanar, aligned ellipses, and (3) the transfer between identical, non-aligned elliptical orbits. Once the above cases have been verified using a GA, other general cases are solved. The GA used is a simple genetic algorithm written in the C programming language and is run on a UNIX operating system.

INTRODUCTION

Several types of transfers between coplanar, non-aligned elliptical orbits have been solved in the past. One of the first and most significant cases of optimal two-impulse transfer was studied in 1925 by Hohmann in the particular case of coplanar, circular orbits. This problem produces an intuitive solution of a transfer ellipse that is tangent to the initial and final orbits at its periapsis and apoapsis. This particular orbit transfer defines what is refered to as the Hohmann transfer [1].

Since the discovery of the Hohmann transfer, many have investigated the case of optimum transfer (minimum total $\Box V$) between elliptical orbits. However, the general equation for this problem is not easily solved for an optimum solution. Instead, the problem was usually simplified to yield an attainable solution. One who has derived these equations fully is Lawden [2]. After deriving them, he simplified the problem into two specific cases. The first is the case in which two elliptical orbits are aligned along their major axis and share a common focus. (Note that because all of the transfer ellipses to be considered in this chapter will always be orbiting the same central body, every ellipse to be discussed shares a common focus.) The solution to this case leads to a trajectory that is similar to the Hohmann transfer, where the transfer ellipse is tangent to the initial and final ellipses at periapsis and apoapsis. A second case is one in which the initial and final elliptical orbits are identical in size, but are non-aligned. This configuration results in the transfer ellipse being tangent to both the initial and final ellipse.

The Hohmann transfer and Lawden's two cases are well defined and are accepted as being correct for coplanar orbits. Because they are straightforward and easily implemented, they will be used as test cases for the GA solution method described in this chapter.

The approach taken in this chapter to define the minimum two-impulse $\Box V$ between coplanar, non-aligned elliptical orbits is done in a general manner. Since this problem is to be solved using a genetic algorithm, the search remains as general as the programmer wishes. Using a GA simply requires the fitness function and the parameters that are to be varied in order to find the best fitness function value. In the case of non-aligned, coplanar elliptical orbits, the variable parameters to be used are eccentricity (e), semi-latus rectum (p), and the orientation angle of the transfer ellipse (ω). Various properties of elliptical orbits are defined in Appendix A. The GA fitness function is related to the total ΔV required to make a transfer between the initial and final orbits, using a transfer ellipse defined by the parameters e, p and ω.

MOTIVATION FOR USING GENETIC ALGORITHMS

Since this problem is dependent on three unknown variables (eccentricity, semi-latus rectum, and angular orientation of the transfer orbit), the search space for the minimum ΔV is three-dimensional. The overriding concern in the selection of an algorithm for optimization is the nature of the fitness function. If the fitness function value is unimodal and easily differentiable, then a gradient method such as Newton's method has significant advantage over others. But the fitness functions encountered in practice are rarely of the above nature. Instead they are discontinuous and jagged. While it is possible to design a gradient-based algorithm to suit a specific problem, there is little incentive for using such methods where simplicity and generality of implementation are needed.

The fitness function for this problem has a rather peculiar nature that prevents the use of gradient-based methods and their variants. Note that every point in three-space with $0 < e < 1$ and positive p corresponds to an elliptical transfer orbit. Not all transfer orbits necessarily intersect the initial and final orbits, implying that, at these points, the fitness function cannot be defined. Although it is desirable that the fitness function be defined everywhere over the variation of the three parameters, in reality, it is defined only over some limited range of the three parameters. The absences of definition of the fitness function over certain ranges may be visualized as "holes" in a graphical representation of the fitness function. That is, if the fitness function were plotted as a surface in four dimensions, that surface would have holes at the points where the function is not defined. Since gradient-based methods use only local information (i.e., they seek an optimum in a neighborhood of the initial guess), it is difficult to use such methods to find the optimal transfer ellipse.

Genetic algorithms are immune to problems created by discontinuity and "holes in domain" such as this, because they use global information. Discontinuities and holes in domain generate "useless" ellipses that are overlooked by the genetic algorithm, while desirable cases – ellipses that intersect both initial and final orbits -- are found. Unlike other methods, the search direction is determined from function evaluation at several randomly generated points. This "population approach" is unique to genetic algorithms and helps avoid the pitfalls of local methods.

FITNESS FUNCTION

The parameters that are required by the fitness function are the orbital parameters stated previously: e, p, and ω. Given these defining properties of an ellipse in a plane, the two ΔV impulses that would be required to make a transfer from the initial to final ellipse, using the specified transfer trajectory, are calculated. The total ΔV required defines the fitness function value and is the value to be minimized by the genetic algorithm. However, because a genetic algorithm maximizes by nature, the minimum total ΔV is calculated by specifying the fitness function in the form of the following equation:

$$\text{Fitness Function} = \text{Constant} - \text{Total } \Delta V. \tag{7.1}$$

This approach will allow the GA to find the minimum ΔV when the fitness function value is maximized.

Once the GA picks values for the three parameters (e, p, and ω), they are passed to the fitness function to be evaluated. Since this ellipse is to be the trajectory that the satellite is to follow while traveling from the initial orbit to the final, it must first be determined that this transfer ellipse intersects both the initial and final orbits. This is done by systematically calculating the location of the points of intersection between the initial and transfer ellipse, and the transfer and final ellipse. If an intersection point does not exist for one of these cases, an orbital transfer using this trajectory is impossible and the fitness function is given the lowest possible value of zero.

In order to determine the intersection points of two orbits, the number of possible intersection points must first be determined. It can be shown that two ellipses sharing a common focus may only have a maximum of two points of intersection. For example, one may say that given the orbit configuration displayed in Figure 7.1, a maximum of four intersection points are possible. However, the orbits illustrated are not possible because the ellipse in the vertical position has a semi-latus rectum that is less than the periapsis. When the trajectory equation [3],

$$r = \frac{p}{1 + e \cos \vartheta} \tag{7.2}$$

is solved at the periapsis location, $\theta = 0°$, the radius at periapsis is given by:

$$r_p = \frac{p}{1 + e \cos 0°} \tag{7.3}$$

By solving (7.3) for p, in terms of eccentricity and radius at periapsis,

$$p = r_p (1+e) \tag{7.4}$$

it can be seen that the semi-latus rectum must always be greater than the radius at periapsis, because eccentricity must always be between 0 and 1 for an ellipse. Therefore, only a maximum of two points of intersection is possible between two ellipses sharing a common focus.

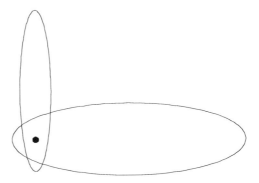

Figure 7.1 - Incorrect representation of orbits.

Finding the exact location of the points of intersection is complicated by the fact that the equation for an ellipse is second order in both the *x* and *y* directions. To solve this problem, an incremental search is used in conjunction with a bi-section root-finder to determine an accurate solution for the intersection points.

Once the intersection points between the initial and transfer ellipse are found, the points are converted from *x* and *y* rectangular coordinates to radius, *r*, and true anomaly, *θ*. Since the origin of the rectangular coordinate frame is at the focus, the radius of the intersection point is given by

$$r = \sqrt{x^2 + y^2}. \tag{7.5}$$

Because the initial and transfer ellipses share a common focus, this radius will be the same for each ellipse. The procedure for finding true anomaly is slightly more involved. First, the direction of the satellite is assumed to be orbiting in a clockwise direction about the focus. From the geometry shown in Figure 7.2, it is seen that the true anomaly (θ, in radians) is given by the following equation:

$$\theta = \pi - \alpha + \omega, \tag{7.6}$$

where α is a reference angle from an arbitrary x-axis and is given by

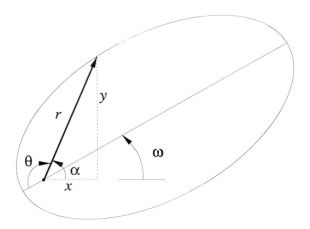

Figure 7.2 - Representation of the true anomaly for a given point of intersection.

Once an intersection point of the two elliptical orbits is found in terms of r and θ, the velocities at both the initial (V_i) and transfer ellipse (V_{tr}) are found at this point using

$$V = \sqrt{\mu\left(\frac{2}{r} - \frac{1}{a}\right)}. \tag{7.8}$$

The flight path angle, γ, for a given velocity is determined by

$$\cos\gamma = \frac{V_n}{V} \tag{7.9}$$

where the normal velocity is given by the equation,

$$V_n = \sqrt{\frac{\mu}{p}}\left(1 + e\cos\theta\right) \tag{7.10}$$

The flight-path angle is determined for both the initial and transfer orbits. Since clockwise orbital motion is assumed, γ will be positive for $0° < \theta < 180°$ and negative for $180° < \theta < 360°$.

The ΔV required for transfer at a particular intersection point is found using the law of cosines for the velocities at that point on the initial and transfer ellipses and the angle between these two velocities. This angle is represented in Figure 7.3. As can be seen, the angle between V_1 and V_{tr} is $\gamma_1 - \gamma_{tr}$, where γ_1 and γ_{tr} have opposite signs. The case where γ_1 and γ_{tr} have the same sign is shown in Figure 7.4. This figure illustrates that when γ_1 and γ_{tr} have the same sign, the angle between V_1 and V_{tr} is once again the difference of the two flight path angles. With this in mind, the change in velocity required to maneuver from the initial orbit to the transfer orbit is

$$\Delta V^2 = V_1^{\,2} + V_{tr}^{\,2} - 2 V_1 V_{tr} \cos(\gamma_1 - \gamma_{tr}). \tag{7.11}$$

Because

$$\cos(-a) = \cos(a), \tag{7.12}$$

the difference of the two flight path angles may be specified in either order, $\gamma_1 - \gamma_{tr}$ or $\gamma_{tr} - \gamma_1$, and the result will be the same.

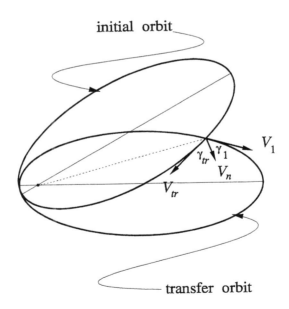

Figure 7.3 - Velocities and flight path angles for two orbits at an intersection point.

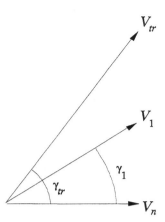

Figure 7.4 - Flight path angles with the same sign.

The above process explains how to determine the ΔV required to maneuver a spacecraft from the initial to the transfer orbit. The process of determining the required ΔV must be performed for each intersection point of the initial and transfer orbit, and for each intersection point of the transfer and final orbit. Then, the minimum ΔV required to go from the initial to the transfer orbit is added to the minimum ΔV required to go from the transfer to the final orbit, resulting in the minimum total ΔV required to transfer from the initial to the final orbit. The fitness function is then found using a modification of equation (7.1), shown below in equation (7.13). For the fitness function, a constant value of 10 TU/DU is used because ΔV is usually of the order of 0 to 5DU/TU. (The canonical units of distance units (DU) and time units (TU) are used throughout this chapter). Making the constant value well above the expected range is important to avoid any negative fitness function values that may arise. Finally, the fitness function is multiplied by a factor (1,000) to give the GA a more substantial number to evaluate. The fitness function is

$$\text{Fitness function} = 1000 \,(10 \text{ DU/TU} - \text{Total } \Delta V). \qquad (7.13)$$

The GA calculates this fitness function for many populations in order to systematically search for the parameters of e, p, and ω that lead to the optimum fitness function value, thus minimizing total ΔV.

TEST CASES

The GA is implemented with the equations shown in the previous section. The specific GA used was developed by Smith, Goldberg, and Earickson [4]. It is a simple genetic algorithm written in the C programming language, and is run on a UNIX operating system. All of these cases in this chapter use populations of 200 and 50 generations. Further details of this particular GA code can be found in [4].

In order to verify the validity of using a GA to solve a problem of this type, several test cases are run which have known solutions. These test cases are the Hohmann transfer [1] and the two cases defined in Lawden [2] – aligned elliptical orbits and non-aligned elliptical orbits of the same size. All three cases are described in more detail below.

Hohmann Transfer

The first test case that is evaluated using the GA is the Hohmann transfer. This transfer is composed of a circular ($e = 0$) initial and final orbit. The specific parameters of the initial and final ellipses are

$$e_i = 0 \; p_i = 1 \, \text{DU} \; \omega_i = 0 \, \text{rad} \qquad (7.14)$$

$$e_f = 0 \; p_{fi} = 1 \, \text{DU} \; \omega = 0 \, \text{rad},$$

where DU refers to the canonical distance unit. Since the GA must be given the set of possible values to try for the parameters, a range for each parameter is specified. For this case, the possible values allowed for the parameters, e, p, and ω, are 0 to 0.99, 0 to 10 DU, and 0 to 360°, respectively. The resulting transfer orbit is shown in Figure 7.5. The initial, transfer and final orbits are denoted by an i, tr, and f, respectively. As can be seen in Figure 7.5, the transfer ellipse is not exactly tangent to the initial and final ellipses as the Hohmann transfer requires. This illustrates the pseudo-random, "brute force" optimization associated with the GA search. Although this solution is not the known best, it is remarkably accurate for being found with such relative ease.

If a more accurate solution is desired for a problem such as this, the GA can be modified slightly to facilitate this. If the range of the best solution can be identified to some "ball park" range, the maximum and minimum values for a parameter can be modified to be this narrower range. By narrowing the range of a parameter, the search space is decreased to include only the most useful parameter values. If this is done successfully, the GA will have an easier task of determining the optimum values because it will have fewer "useless" values to process.

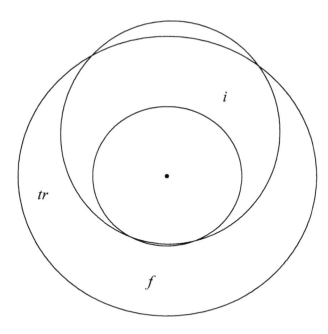

$$e_{tr} = 0.391 \quad p_{tr} = 1.353 \text{DU} \quad \omega_{tr} = 1.519 \text{rad} \quad \Delta V_{total} = 0.44128 \frac{\text{DU}}{\text{TU}}$$

Figure 7.5 - Result for the initial computation of the transfer orbit for circular initial and final orbits.

The narrowing of the parameter range is done for each of the test cases to check the results found using the GA. For the Hohmann transfer case, the ranges of e and p are narrowed to 0.2 to 0.5 and 1 to 2 DU, respectively. The range of ω is reduced drastically to 0 to 0.1 radians. This large reduction in range is justified because ω for the Hohmann transfer case carries no meaning, due to the fact that the initial and final orbits are circular. After these parameter ranges are narrowed, the resulting transfer orbit is shown in Figure 7.6.

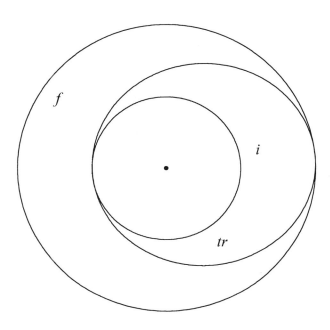

$$e_{tr} = 0.336 \, p_{tr} = 1.330 \text{DU} \, \omega_{tr} = 0.086 \text{rad} \, \Delta V_{total} = 0.29335 \, \tfrac{\text{DU}}{\text{TU}}$$

Figure 7.6 - Transfer orbit result for circular initial and final orbits, after narrowing parameters.

This figure displays, almost exactly, the Hohmann transfer solution. Between the solution shown in Figure 7.5 and the one shown in Figure 7.6, the V requirements are reduced by 34% as a result of reducing the list of possible parameters to be used by the GA.

Transfer between Aligned Ellipses

The second case is one described in Lawden [2]. For this test case, the parameters specifying the initial and final ellipses are

$$e_i = 0.8 \; p_i = 1 \, \text{DU} \; \omega_i = 0 \, \text{rad} \qquad (7.15)$$
$$e_i = 0.8 \; p_i = 1 \, \text{DU} \; \omega_i = 0 \, \text{rad} \; .$$

For the first implementation of the GA, the parameter ranges are 0 to 0.99 for eccentricity, 0 to 10 DU for the semi-latus rectum, and 0 to 2π for the orientation angle. After the initial run of the genetic algorithm, the optimum orbit is shown in Figure 7.7. Then, the parameter ranges are narrowed to 0.5 to 0.99 for *e*, 1 to 2 DU for *p*, and 0 to 0.1 radians for ω, in order to obtain a more accurate solution. Figure 7.8 displays the transfer ellipse obtained from the reduction of the parameter ranges.

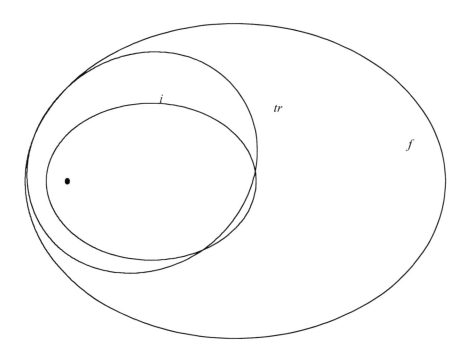

$$e_{tr} = .6586 \quad p_{tr} = 1.7403\text{DU} \quad \omega_{tr} = .1979\text{rad} \quad \Delta V_{total} = 0.16025 \tfrac{\text{DU}}{\text{TU}}$$

Figure 7.7 - Initial solution for aligned orbits.

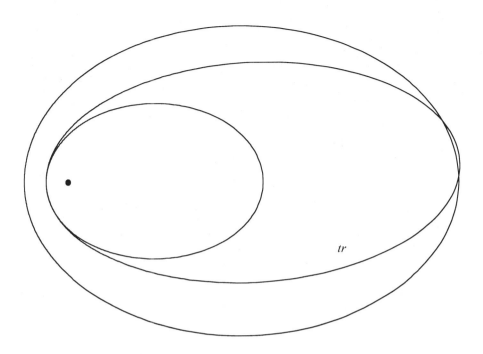

$$e_{tr} = 0.895 \, p_{tr} = 1.053 \text{DU} \, \omega_{tr} = 0.037 \text{rad} \Delta V_{total} = 0.10136 \tfrac{\text{DU}}{\text{TU}}$$

Figure 7.8 - Transfer orbit solution for aligned elliptical orbits, after the parameter ranges have been reduced.

For this case, the optimum transfer as described in Lawden is found. By narrowing the parameter ranges, the required $\Box V$ is reduced by 37%. This reduction of ΔV is expected because Figure 7.8 shows a transfer orbit that is more tangential to the initial and final orbits and is more elliptical in nature.

Non-Aligned, Identical Elliptical Orbits

The third test case is to find the transfer orbit between two non-aligned elliptical orbits of the same size. The parameters specifying the initial and final ellipses of this case are

$$e_i = 0.7 \ \ p_i = 1 \, \text{DU} \ \ \omega_i = 0 \, \text{rad}$$
$$e_i = 0.7 \, p_i = 1 \, \text{DU} \, \omega_i = 0 \, \text{rad} \tag{7.16}$$

After the first pass through the genetic algorithm, with the initial parameter ranges once again set to 0 to 0.99 for eccentricity, 0 to 10 DU for the semi-latus rectum, and 0 to 2π for the orientation angle, the transfer orbit solution is shown in Figure 7.9. After the parameter ranges are narrowed to 0.4 to 0.99 for e, 1 to 3 DU for p, and 0 to 1.047 radians for ω, the solution is shown in Figure 7.10.

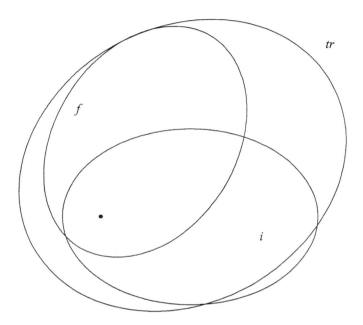

$$e_{tr} = 0.569 \quad p_{tr} = 1.784\text{DU} \quad \omega_{tr} = 0.575\text{rad} \quad \Delta V_{total} = 0.29774\,\tfrac{\text{DU}}{\text{TU}}$$

Figure 7.9 - Initial solution for the case of non-aligned elliptical orbits of the same size.

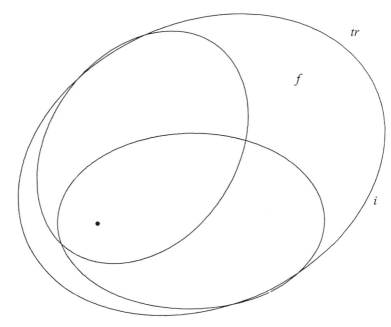

$$e_{tr} = 0.628 \qquad p_{tr} = 1.722 \, \text{DU} \qquad \omega_{tr} = 0.522 \, \text{rad} \qquad \Delta V_{total} = 0.274$$

Figure 7.10 - Solution for the transfer between two non-aligned elliptical orbits of the same size, after the parameter ranges have been reduced.

Here, once again, the resulting transfer orbit is very nearly tangential to both the initial and final orbits. The narrowing of the possible parameters to be used by the GA, reduced the required ΔV by 8%.

Based on the results displayed for the three test cases, the GA successfully identified the known solutions for each of the cases and did this with relative ease. Now, the stage is set to find the optimum $\Box V$ for cases in which the solution has not previously been proven.

RESULTS

As with the test cases, each of these cases are run once to find a near optimum solution. For each of the initial runs of the genetic algorithm, the parameter ranges will remain 0 to 0.99 for eccentricity, 0 to 10 DU for semi-latus rectum, and 0 to 2π radians for the orientation angle. Then, the parameter ranges are narrowed to yield an answer that is closer to an exact solution.

Case 1

The first case to be attempted is a case with low eccentricity values. The orbit parameters used are

$$e_i = 0.2 \; p_i = 1.5 \, \text{DU} \; \omega_i = 0.3 \, \text{rad} \qquad\qquad (7.17)$$

$$e_f = 0.2 \; p_f = 2.5 \, \text{DU} \; \omega_{fi} = 2.8 \, \text{rad}$$

The initial implementation of the GA yields the solution illustrated in Figure 7.11. After narrowing the parameter ranges to 0.2 to 0.8 for e, 1 to 3 DU for p, and the range for ω remains 0 to 2π, the solution is slightly improved to the results shown in Figure 7.12.

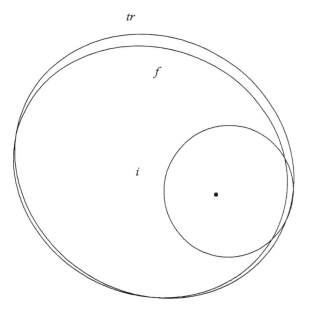

$$e_{tr}' = 0.477 \quad p_{tr} = 2.640 \, \text{DU} \quad \omega_{tr} = 2.705 \, \text{rad} \quad \Delta V_{total} = 0.25478 \, \tfrac{\text{DU}}{\text{TU}}$$

Figure 7.11 - Initial solution for Case 1.

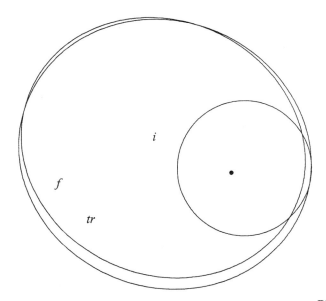

$$e_{tr} = 0.472 \quad p_{tr} = 2.658 \, \text{DU} \quad \omega_{tr} = 2.714 \, \text{rad} \quad \Delta V_{total} = 0.24357 \frac{\text{DU}}{\text{TU}}$$

Figure 7.12 - Solution for Case 1 after the parameter ranges are narrowed.

In this case, reducing the ranges of parameters improved the required ΔV only by 4%. However, both figures show very similar transfer orbits that are both nearly tangent to the initial and final orbits, thus implying that both the initial and final solutions are near optimal.

Case 2

The second case uses one ellipse that is highly elliptical and one that is slightly more circular. The parameters of these ellipses are

$$e_i = 0.6 \, p_i = 1 \, \text{DU} \, \omega_i = 0.578 \, \text{rad} \tag{7.18}$$
$$e_f = 0.9 \, p_f = 2 \, \text{DU} \, \omega_f = 1.047 \, \text{rad}$$

The resulting transfer ellipse from the initial implementation of the GA is shown in Figure 7.13. After the parameters space is narrowed to 0.5 to 0.99 for e, 1 to 2 DU for p, and 0 to 1.047 radians for ω, the improved transfer ellipse is shown in Figure 7.14.

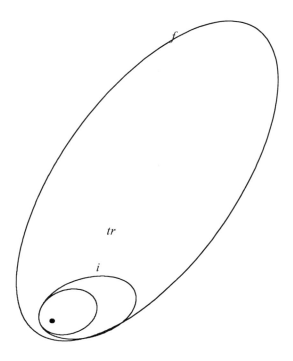

$$e_{tr} = 0.753 \quad p_{tr} = 1.120\text{DU} \quad \omega_{tr} = 0.414\text{rad} \quad \Delta V_{total} = 0.29653\,\tfrac{\text{DU}}{\text{TU}}$$

Figure 7.13 - Initial solution for Case 2.

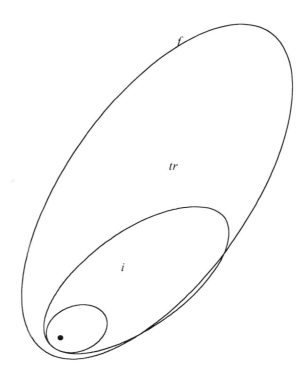

$$e_{tr} = 0.8853\, p_{tr} = 1.1980\,\text{DU}\,\omega_{tr} = 0.7174\,\text{rad}\,\Delta V_{total} = 0.23258\,\tfrac{\text{DU}}{\text{TU}}$$

Figure 7.14 - Solution for Case 2, after the parameter ranges have been narrowed.

For Case 2, reducing the parameter list results in a 22% reduction in the ΔV requirement. The reduction in ΔV is expected since the transfer ellipse is larger and more eccentric, thus allowing for a smoother velocity transition from the transfer to the final ellipse.

Case 3

The final orbits to be considered are both highly elliptical and are separated by almost $90°$. The parameters for the initial and final ellipses are

$$e_i = 0.8 \; p_i = 1.3 \text{ DU} \; \omega_i = 0.578 \text{ rad} \tag{7.19}$$
$$e_f = 0.9 \; p_f = 1 \text{ DU} \; \omega_f = 2.047 \text{ rad}$$

Figure 7.15 displays the initial solution for the transfer ellipse. After refining the range of the parameters to 0.5 to 0.99 for e, 3 to 6 DU for p, and 0.578 to 2.047 radians for ω, the result improves, as shown in Figure 7.16.

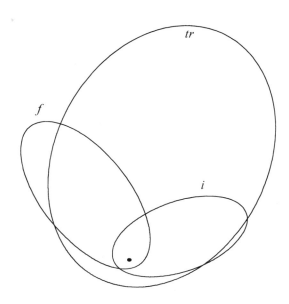

$$e_{tr} = .798 \quad p_{tr} = 3.174\text{DU} \quad \omega_{tr} = 1.356\text{rad} \quad Delta\, V_{total} = 0.36995\, \tfrac{\text{DU}}{\text{TU}}$$

Figure 7.15 - Initial transfer ellipse solution for Case 3.

After reduction of the allowable parameter ranges, the resulting transfer ellipse is larger, more eccentric, and requires 5% less ΔV than the case shown in Figure 7.15. However, the transfer orbit is not as tangential to the final ellipse as the cases previously shown. This case of two highly elliptical orbits separated by approximately 90° illustrates more of a challenge for the genetic algorithm. In a case such as this, many of the transfer orbits contained in a population that tangentially intersect the final orbit are likely to not intersect the initial orbit at all, rendering it useless. However, it should be noted that the solution in this case does come somewhat near to a tangential intersection, and this result was found with relative ease.

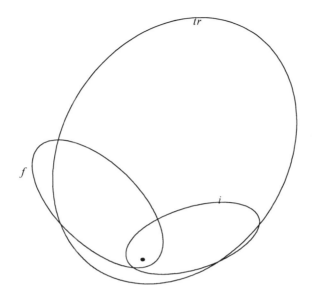

$$e_{tr} = .822 \quad p_{tr} = 3.257 \text{DU} \quad \omega_{tr} = 1.377 \text{rad} \quad Delta\, V_{total} = 0.35034 \tfrac{\text{DU}}{\text{TU}}$$

Figure 7.16 - Solution for the transfer ellipse for Case 3, after the parameter ranges have been narrowed.

CONCLUSIONS

From the results of the three test cases, it can be seen that the GA is getting near-exact solutions for these known cases. Limiting the ranges for the parameters did prove, in every case, to lower the amount of fuel required to make the orbit transfer. Therefore, if the user could provides the genetic algorithm with ranges of the parameters that are more likely to produce a good transfer orbit, better performance will result.

In addition, the results of each of the cases agree very closely with the intuitive solution for the transfer ellipse because all of the resulting transfer orbits shown are nearly tangent to both the initial and final ellipse, where Case 3 may prove to deviate the farthest from what is intuitively expected. In Case 3, the difficulty for the GA to find a solution that is more tangential to the final orbit may be due to the fact that the orbits are separated by almost 90° and are highly elliptical. Overall, the accuracy of the results and the ease with which they were found do nothing but further verify the use of a GA for such a problem.

For all of the cases, the GA proves to be a desirable method in which to find transfer-orbit solutions, because this type of solution method requires no derivative information for the complicated equations that describe this type of two-impulse orbital transfer. In order to find a solution, the GA only requires the straightforward equations that define the ΔV needed to transfer through some arbitrary transfer orbit, instead of complicated gradient information.

The successful performance of the genetic algorithm for the application demonstrated here, suggests that the minimum fuel for orbital transfer problem may be explored further using the GA. It may be applied in more realistic cases, such as multi-impulsive maneuvers and orbital transfers between non-coplanar orbits.

REFERENCES

[1] Hohmann, W. (1960). *Die Ereichbarkeit der Himmelskorper* (The Attainability of Heavenly Bodies), NASA, Technical Translations F-44.

[2] Lawden, D. S. (1962). Impulsive transfer between elliptical orbits. *Optimization Techniques*, edited by G. Leitmann, Academic, New York, Chapter 11.

[3] Bate, R. R., Mueller, D. D., White, J, E. (1971). *Fundamentals of Astrodynamics*. New York, NY: Dover Publications, Inc.

[4] Smith, R. E., Goldberg, D. E., Earickson, J. A. (1991). SGA-C: A C-language implementation of a simple genetic algorithm. *TCGA Report No. 91002*.

APPENDIX A

Properties of an Elliptical Orbit

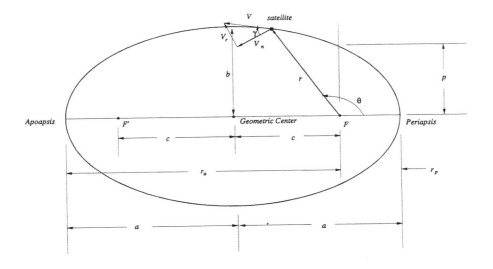

Figure 7.A1 - Various properties of an elliptical orbit.

a   =   semimajor axis = $\dfrac{(r_a + r_p)}{2}$

b   =   semiminor axis

e   =   eccentricity = $\dfrac{(r_a - r_p)}{(r_a + r_p)}$

c   =   ea

θ =   true anomoly

r_a =   apoapsis radius = a (1 + e)

r_p =   periapsis radius = a (1 – e)

p   =   semilatus rectum = $a(1 - e^2) = \dfrac{b^2}{a} = r_p(1+p) = r_a(1-e)$

γ =   flight-path angle

F   =   prime focus – location of the central attracting body

F'   =   empty focus

V   =   satellite velocity

V_n =   normal satellite velocity

V_r =   radial satellite velocity

CHAPTER 8

OPTIMIZED NON-COPLANAR ORBITAL TRANSFERS USING GENETIC ALGORITHMS

Dedra C. Moore
KBM Enterprises, Inc.
Huntsville, AL 35814
e-mail: dmoore@kbm-inc.com

ABSTRACT

The problem of optimizing the transfer orbit between an initial orbit and a desired final orbit is not easily solved. The governing equations are difficult to solve in closed form. Additionally, the problem is even more difficult when the initial and final orbits are not coplanar. This report examines the use of a genetic algorithm (GA) to find the transfer orbit and plane change that minimizes the total velocity change (ΔV_{TOT}) needed. Cases where the initial and final orbits are both circular and elliptical are investigated. Accuracy of the GA is verified by both coplanar and non-coplanar Hohmann transfers between circular orbits, for which analytical solutions exist.

LIST OF SYMBOLS

ΔV_{TOT}	Total velocity change
ΔV	Velocity change
e_t	Eccentricity of transfer orbit
e	Eccentricity
ω	Argument of perigee
a	Semi-major axis
i	Inclination measured from the equatorial plane
Ω	Longitude of the ascending node
α_1	Initial plane change done in conjunction with the first velocity change
ν	True anomaly angle
r_p	Perigee radius
V_p	Velocity at perigee

INTRODUCTION AND BACKGROUND INFORMATION

In many instances, objects orbiting the earth are not initially in their desired orbits. For example, satellites placed in a low earth orbit by the space shuttle may require a higher altitude to provide maximum efficiency. Satellites also often drift over time from their original paths due to orbital perturbations. In both cases, an orbital transfer must take place to correct the problem. For many years, scientists who work in the field of orbital mechanics have been interested in finding the minimum velocity change to get from one orbit to another. Generally, this concern stems from the need to limit the weight and amount of fuel needed on board. However, in closed form, the governing equations which optimize the transfer are difficult to solve. Typically, assumptions have been made to elucidate the solution process.

The best known and simplest transfers was developed by Hohmann [1]. He examined coplanar circular orbits and determined that the optimum transfer orbit was an elliptical orbit whose perigee point lay on the initial orbit and apogee point lay on the destination orbit. The maneuver is executed by applying a tangential ΔV at the departure point on the initial orbit and another tangential ΔV to circularize into the final orbit (Figure 8.1). This procedure can be solved analytically. Hohmann also developed a procedure for finding an optimum transfer for circular non-coplanar orbits. This maneuver is similar to the coplanar transfer in that the transfer orbit perigee and apogee points lie on the initial and final orbits, and velocity changes are performed at these locations. However, a plane change must also be incorporated into one or both velocity changes. It is the optimization of the velocity change that makes this problem difficult to solve analytically. The governing equation cannot be written in terms of the plane change angle; it must be solved iteratively.

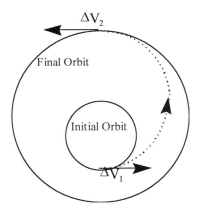

Figure 8.1 - Classic Hohmann transfer.

In 1962, Bender [2] published a paper which expanded Hohmann's work to analyze elliptical orbits. Coplanar orbits were still assumed, however his work included the possibility of elliptic initial and final orbits. This study was restricted to two classes of transfer orbits – 180° transfer and cotangential transfer. The 180° transfer was defined as a transfer where the arrival and departure points are separated by 180 degrees and not necessarily at the apses of either orbit (Figure 8.2).

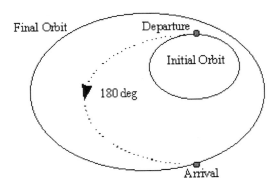

Figure 8.2 - 180° transfer case.

For the cotangential transfer case, the initial and final orbits were assumed to be coplanar (Figure 8.3). For each of these orbit types, once the departure point is defined, the transfer orbit is fixed. Therefore, each case was optimized by iteratively varying the location of the first ΔV or departure point.

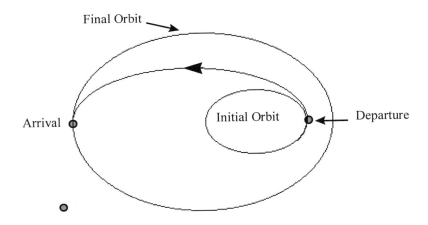

Figure 8.3 - Cotangential transfer case.

Lawden [3] studied coplanar cases similar to the work of Bender, but also examined two cases with elliptic initial and final orbits. The first case involved elliptic orbits aligned along their major axis. He concluded that the optimum transfer is tangential at its apses to the initial and final orbits, and its axis is aligned with the other two axes. In this situation, the optimum transfer orbit is defined based on the apses location of the two given orbits and requires no iteration. The second case examined by Lawden involved initial and final orbits of identical size, but with skewed major axes (Figure 8.4). This is analogous to the cotangential case examined by Bender, where the transfer orbit is tangential to both given orbits and optimized by iteratively picking a departure point.

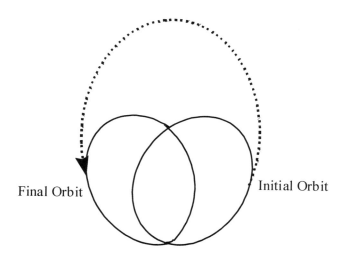

Figure 8.4 - Transfer between equal ellipses.

In 1966, Baker [4] examined the possibility of executing three-dimensional orbit transfers with a bi-elliptic transfer. This maneuver involved three velocity changes. The first and third were performed at the departure of the initial orbit and the arrival at the final orbit. The second is a velocity change at the apogee point on the initial transfer orbit onto another intermediate transfer orbit. Baker's study was limited to both circular initial and final orbits and examined optimally, splitting the plane change required between the three ΔV's. The governing equation which defines the total velocity change needed is written in terms of each of the three plane changes and cannot be solved for either angle. Here again, an iterative method is required to find the three angles.

Reichert [5] used a GA to minimize the total ΔV in three coplanar orbit transfer cases. The first was the general transfer between circular orbits studied by Hohmann in 1925, which can typically be solved by hand. The second and third were the two cases studied by Lawden. The classical solution to these cases was used to verify the accuracy of the genetic algorithm.

In 1995, Pinon and Fowler [6] used a GA to generate two-dimensional lunar trajectories. The goal was to analyze orbital transfers from the earth to the moon. Their results were comparable to the trajectories used by the Apollo missions. They concluded that the two-dimensional trajectories produced by the GA could be used as a starting point for three-dimensional transfers.

The above-mentioned studies were simplified to have either coplanar or circular initial and final orbits or both. The purpose of the current effort is to examine the use of a GA in non-coplanar orbit transfers. A GA is effective in finding near optimum solutions, but does not guarantee to find the optimum solution. Because of the GA's capability, test cases will include both circular and elliptical problems. This situation is encountered more often in orbital maneuvers. However, this problem, like the coplanar transfer, is not easily solved. In this case, not only must an optimum transfer orbit be found, but also the amount of plane change at each ΔV.

PROBLEM STATEMENT

In this chapter, orbital perturbations will be ignored, impulsive maneuvers will be assumed, and test cases will be limited to bodies orbiting the earth. In a real world analysis, these issues would need to be addressed. However, neglecting them will be sufficient for this study. This investigation will limit the number of velocity changes to two, the first when leaving the initial orbit and the second at the intersection of the transfer orbit and the final orbit. It will also be assumed that the semi-major axes of the initial, final and transfer orbits are aligned with one another.

The analysis will begin by user specification of the initial and final orbits, which will be limited to elliptical or circular. Parameters used to define an orbit will be the eccentricity (e), the argument of perigee (ω), the semi-major axis (a), the inclination with respect to the equatorial plane (i), and the longitude of the ascending node (Ω). The GA will be used to search for defining parameters for the transfer orbit and an initial plane change (α_1) that minimizes the velocity change needed.

This study assumes the first velocity change will occur at the perigee point of the initial orbit and of the transfer orbit. Therefore, the perigee radius of the initial and transfer orbits will correspond. Since the perigee radius of the transfer orbit is defined by the initial orbit, the semi-major axis can be calculated (Equation 8.1) and need not be one of the search parameters for the GA.

$$a_t = \frac{r_p}{1 - e_t} \tag{8.1}$$

where

$$r_p = a(1 - e) \tag{8.2}$$

The perigee velocity on both the initial and transfer orbits can then be calculated using Equation 8.3.

$$V_p = \sqrt{\mu\left(\frac{2}{r_p} - \frac{1}{a}\right)} \tag{8.3}$$

where

$$\mu = 3.986 * 10^5 \frac{km^3}{\sec^2} \tag{8.4}$$

The velocity change required to go from the initial orbit to the transfer orbit (ΔV_1) can be found by applying the law of cosines (Figure 8.5, Equation 8.5).

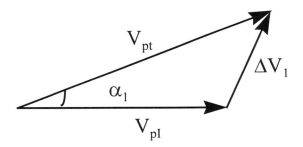

Figure 8.5 - Physical representation of the law of cosines at the first velocity change.

$$\Delta V_1 = \sqrt{V_{pt}^2 + V_{pl}^2 - 2V_{pt}V_{pl}\cos(\alpha_1)} \tag{8.5}$$

where

V_{pt} = perigee velocity on the transfer orbit
V_{pl} = perigee velocity on the initial orbit

In order for the transfer orbit to be sufficient, it must intersect the final orbit (Figure 8.6). The next step will be to determine the intersection location. The initial, random population will undoubtedly contain transfer orbits that do not intersect the final orbit and are not possible solutions. In these situations, the objective function will be penalized according to how badly the transfer orbit misses the final orbit (details will be discussed later).

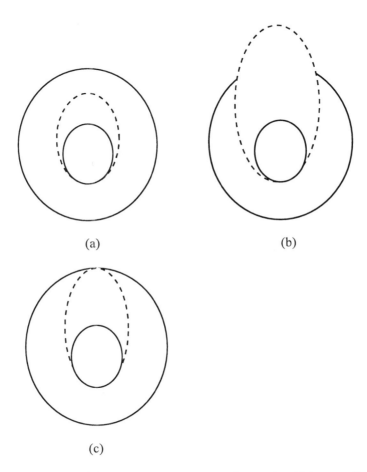

(a)   (b)

(c)

Figure 8.6 - (a) transfer orbit does not intersect final orbit, (b) transfer orbit intersects final orbit at two locations, (c) transfer orbit intersects final orbit at one location.

The process of finding the intersection point or points of two non-coplanar elliptical orbits is not a trivial task. The point will be defined by the following two values: a true anomaly angle on the transfer orbit, v_t, and a true anomaly angle on the final orbit, v_f (Figure 8.7).

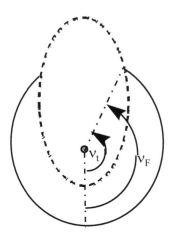

Figure 8.7 - Representation of intersection points.

The first step is to represent the radius vector to the intersection point on both the transfer orbit and the final orbit in a "perifocal coordinate system" (Figure 8.8). The perifocal coordinate system is defined as having a fundamental plane in the plane of the orbit. Positive x points toward perigee of the orbit and positive y is rotated 90° in the direction of the orbital motion. This is one of the most convenient coordinate systems for describing an orbit because the z component is always zero (Equation 8.6).

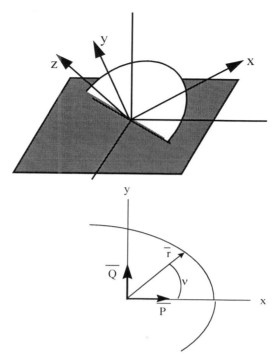

Figure 8.8 - Perifocal coordinate system.

$$\vec{r} = \begin{bmatrix} r\cos(v) \\ r\sin(v) \\ 0 \end{bmatrix} \qquad (8.6)$$

where

$$r = \frac{a(1 - e^2)}{1 + e\cos(v)} \qquad (8.7)$$

The second step is to use a coordinate transformation to obtain a representation of each orbit in a common reference frame. The common coordinate system is the "geocentric equatorial coordinate system." This system has its origin at the earth's center. The fundamental plane is the earth's equatorial plane and the positive x-axis points in the vernal equinox direction. The z-axis points in the direction of the north pole (Figure 8.9). The transformation [7] between the two coordinate systems is done by use of a rotation matrix, R (Equation 8.8).

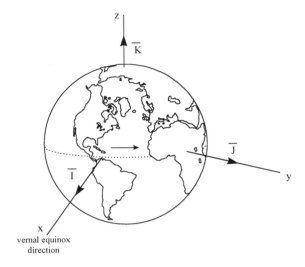

Figure 8.9 - Geocentric equatorial coordinate system.

$$
\begin{bmatrix} r_I \\ r_J \\ r_K \end{bmatrix} = R \begin{bmatrix} r_P \\ r_Q \\ r_W = 0 \end{bmatrix} \tag{8.8}
$$

The point of intersection occurs when each coordinate in the geocentric equatorial coordinate system of the transfer equals that of the final orbit. This system of nonlinear equations can be solved for the variables, v_t and v_f, at the point of intersection using a Taylor Series approach [8].

Once the location of intersection has been found, the radius to the intersection can be determined using the conic equation (Equation 8.7). The velocity on each orbit at the intersection point can then be found using Equation 8.9.

$$
V = \sqrt{\mu\left(\frac{2}{r} - \frac{1}{a}\right)} \tag{8.9}
$$

The velocity change required to go from the transfer orbit to the final orbit at the intersection location can then be calculated by applying the law of cosines (Figure 8.10, Equation 8.10).

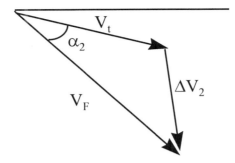

Figure 8.10 - Physical representation of the law of cosines at the second velocity change.

$$\Delta V_2 = \sqrt{V_F^2 + V_t^2 - 2V_t V_F \cos(\alpha_2)} \qquad (8.10)$$

where

V_t = velocity on the transfer orbit at intersection point
V_F = velocity on the final orbit at intersection point

$$\alpha_2 = \left| i_1 - i_2 \right| - \alpha_1 \qquad (8.11)$$

where

i_1 = the inclination of the initial orbit measured from the equatorial plane
i_2 = the inclination of the final orbit measured from the equatorial plane

Hence, the total ΔV is:

$$\Delta V_{TOT} = \left| \Delta V_1 \right| + \left| \Delta V_2 \right| \qquad (8.12)$$

It should be noted that for those transfer orbits which intersect the final ellipse twice, two values for ΔV_{TOT} will exist. However, since the semi-major axes were assumed to be aligned, the magnitude of each ΔV_{TOT} is equal.

GENETIC ALGORITHM PARTICULARS

Fitness Function

Since the goal is to determine a transfer orbit which minimizes the total ΔV needed to go from an initial orbit to a final orbit, the fitness value is considered to be the value of Equation 8.12 plus a penalty for those orbits which do not cross the final orbit (Equation 8.13).

$$Fitness = \Delta V \left[1 + k \left(\frac{\Delta r_a}{r_{a\_final}} \right) \right] \qquad (8.13)$$

where

k = a weighting factor of 1500 (determined by numerical experimentation)
r_{a_final} = the radius of apogee for the final orbit
Δr_a = the distance the proposed transfer orbit misses the final orbit (Figure 8.11)

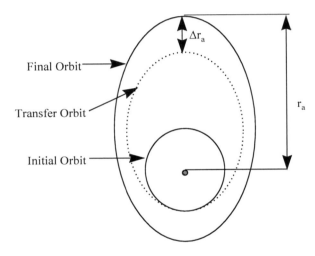

Figure 8.11 - Diagram of how penalty function is assessed when the transfer orbit does not intersect the final orbit.

The penalty is added, instead of setting the fitness values to a large number, to ensure that any good genetic material which might exist in these solutions has a chance to survive to the next generation.

Coding Scheme

As described above, the goal of this project is to demonstrate the utility of using a genetic algorithm to find a transfer orbit to minimize the total velocity change needed to go from an initial orbit to a final orbit. The transfer orbit will be defined by the eccentricity (e), the argument of periapsis (ω), the longitude of the ascending node (Ω), and the initial plane change ($α_1$). The length of the binary string which represents these four parameters is found by specifying the accuracy of each parameter and finding the necessary substring length, and then concatenating (Equation 8.14).

$$l = \frac{\log\left(\dfrac{U_{MAX} - U_{MIN}}{\pi} + 1\right)}{\log(k)} \tag{8.14}$$

where

π = the accuracy of the parameter
$k = 2$ for a binary representation
l = length of substring
U_{max} = maximum value the parameter can take
U_{min} = minimum value the parameter can take

The following table shows the accuracy chosen for each parameter and the substring length needed based on solving Equation 8.14.

Table 8.1 – Substring length for each parameter based on chosen accuracy and maximum and minimum values.

Parameter	Accuracy	U_{min}	U_{max}	Substring Length
e	0.001	0	1	10
ω	0.001	0	2π radians	13
Ω	0.001	0	2π radians	13

The substring length for α_1 is not easily calculated. It must range between the inclination of the initial and final orbits which vary from problem to problem. Therefore, a set length of fifteen was chosen for this parameter. If the substring lengths of each of the four parameters are concatenated, the total string length is 51.

The string is then arranged with the eccentricity and plane change parameter representations adjacent to one another. The eccentricity is important in defining the shape of the transfer orbit, and the plane change angle is significant in determining the amount of plane change which must be transversed at each velocity change. They are placed adjacent to one another to reduce the likelihood of destroying good combinations of these two parameters by crossover. Therefore, the binary string is set up as follows:

Table 8.2 - Position of each parameter in string.

Parameter	E	α_1	ω	Ω
Position in string	1-10	11-25	26-38	39-51

Once the string is defined, an initial population of strings is randomly generated. The fitness of each string in the population is found by evaluating Equation 8.13. Then, reproduction using tournament selection, two-point crossover, and mutation are used to generate subsequent generations and search for acceptable values for e, ω and Ω of the transfer orbit and α_1 which minimize Equation 8.13.

Tournament selection is executed by randomly picking two strings from the current population, and comparing their fitness values. The string with the lower fitness value is automatically put into the mating pool for the new population. Two-point crossover is accomplished by randomly picking two strings from the mating pool, then randomly picking two locations in the string length, based on a probability of crossover of 0.9, crossing the strings at the first location, and then crossing the strings at the second location. A mutation operator with a probability of 0.001 is used to introduce new genetic material into the population. If this operator is enacted on a bit in the string, it will be flipped. In other words, a "1" in a bit position is replaced with a "0," and a "0" with a "1" [9]. Once a new population of strings is generated, the fitness values of each string are found. In this investigation, this procedure was used to generate ten generations of populations.

RESULTS

The ability of the GA to locate optimum orbital transfers can be tested by comparing its results to some well-known results. If the initial and final orbits are circular then the genetic algorithm should produce a transfer orbit equal to or near a Hohmann transfer, which is the minimum energy transfer between two coplanar circular orbits (Figure 8.12). Because the analytical solution to this situation is known, it was used to find a value for the weighting factor "k" in Equation 8.13 which produced a transfer near the known optimum.

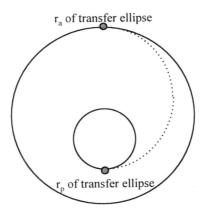

r_a of transfer ellipse

r_p of transfer ellipse

Figure 8.12 - Hohmann transfer between two circular orbits.

For the first test case, two coplanar circular orbits were used. The initial orbit had a semi-major axis of 7,000 km and the final orbit had a semi-major axis of 7,500 km. When the genetic algorithm was run and compared to the known analytical solution, the following results were obtained (Table 8.3):

Table 8.3 - Comparison of GA to analytical solution for transfer orbit for co-planar circular initial and final orbits. (GA population size=80, number of generations=10, probability of crossover=0.9, probability of mutation=0.001).

	Analytical	**GA**
Eccentricity	0.03448	0.03519
α_1	0 deg	0 deg
ω_t	0 deg	23.6 deg
Ω_t	0 deg	243 deg
ΔV_{TOT}	0.2558 km/sec	0.2661 km/sec

The transfer orbit resulting from the GA is slightly more elliptical than the optimum Hohmann transfer orbit (Figure 8.13). However, genetic algorithms do not promise to locate optimum solutions, only near optimum solutions. If this is considered, the GA performed quite well in this situation, as indicated in the fact that the ΔV it determined was only 4.02% larger than the optimum value.

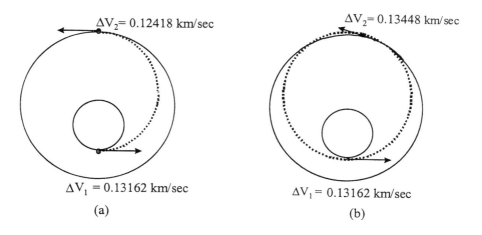

Figure 8.13 - (a) Optimum transfer orbit (Hohmann transfer), (b) GA produced transfer orbit.

The next test case is a non-coplanar Hohmann transfer. If the inclination between the initial and final orbits is set to zero, the total ΔV needed can be found using Equations 8.7, 8.9 and 8.12. For transfers between circular orbits where the inclination between them is not zero, the accepted solution method is to solve Equation 8.15 which is derived by applying the law of cosines at the time of ΔV_1 and ΔV_2, iteratively for the optimum α_1.

$$\frac{\partial \Delta V_T}{\partial \alpha_1} = \frac{V_{P_{TR}} V_{c1} \sin(\alpha_1)}{\Delta V_1} - \frac{V_{A_{TR}} V_2 \sin(|i_1 - i_2| - \alpha_1)}{\Delta V_2} \qquad (8.15)$$

If the same circular orbits are used again, however, with an initial inclination of 5 deg and a final inclination of 10 deg, the following comparison results are obtained (Table 8.4):

Table 8.4 - Comparison of GA to analytical solution for transfer orbit for non-planar circular initial and final orbits. (GA population size=80, number of generations=10, probability of crossover=0.9, probability of mutation=0.001.)

	Analytical	**GA**
Eccentricity	0.03448	0.03499
α_1	0.4869 deg	0.4890 deg
$\Delta V_{\square\square\square}$	1.349 km/sec	1.351 km/sec

The transfer orbit produced by the GA is slightly more elliptical than the analytical optimum (Figure 8.14). However, the ΔV is only 0.148% greater than the optimum value.

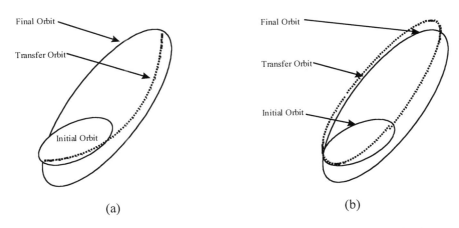

Figure 8.14 - (a) Optimum transfer orbit, (b) GA produced transfer orbit for non-coplanar circular initial and final orbits.

The GA was further tested with the non-coplanar Hohmann case by applying it to a range of final orbit inclinations. Circular initial and final orbits were used with respective semi-major axes of 7,500 km and 8,000 km. The final orbit inclination was varied between $0°$ and $90°$. The total ΔV's obtained from the GA produced orbits were then plotted with the analytical solutions (Figure 8.15). The GA produced results were comparable to the analytical for each inclination.

Final Orbit Inclination vs. Total DV

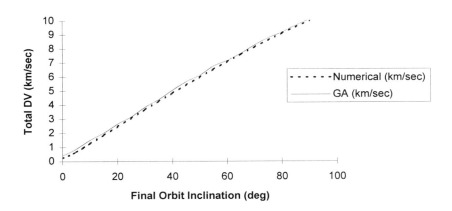

Figure 8.15 - Total ΔV produced by numerical and GA methods versus final orbit inclination for a non-coplanar Hohmann transfer (GA population size = 150, number of generations = 10, probability of crossover = 0.9, probability of mutation = 0.001).

Once of the GA was validated by comparison to analytical solutions, more complex problems were addressed; problems which do not have known analytical solutions. The following cases were used to apply the GA to non-coplanar elliptical orbits:

Case I

Table 8.5 - Initial and final orbits parameters for two non-coplanar elliptical orbits.

	Initial Orbit	**Final Orbit**
Eccentricity	0.1	0.2
Inclination	0°	10°
Semi-major axis	6500 km	6800 km

Number of Generations vs. Total DV

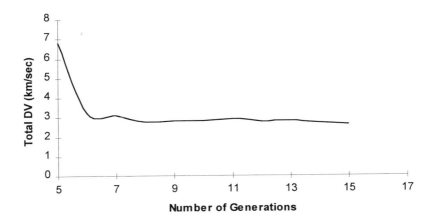

Figure 8.16 – Total $\Box$V produced by the GA vs. number of generations.

When the GA is run on this case, the results shown in Table 8.6 were obtained. These results are also shown in Figure 8.17.

Table 8.6 - GA results for transfer orbit parameters for non-planar elliptical initial and final orbits. (GA population size=150, number of generations=15, probability of crossover=0.9, probability of mutation=0.001)

	GA Results
Eccentricity	0.1652
α_1	0.6222 °
Semi-major axis	7007.6 km
$\Delta V_{\Box\Box\Box}$	2.6019 km/sec

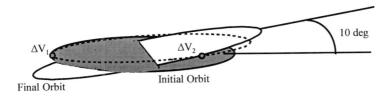

Figure 8.17 - Initial and final orbits along with GA-produced transfer orbit.

Case II

Table 8.7 - Initial and final orbits parameters for two non-coplanar elliptical orbits.

	Initial Orbit	**Final Orbit**
Eccentricity	0.01	0.5
Inclination	5°	12°
Semi-major axis	6500 km	7500 km

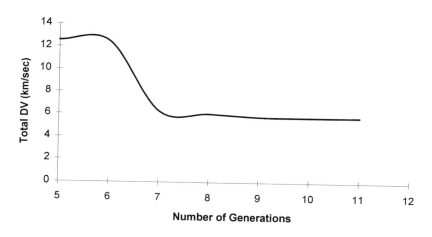

Number of Generations vs. Total DV

Figure 8.18 – Total ΔV produced by the GA vs. number of generations.

When the GA is run on this case, the following results (Table 8.8) (Figure 8.19), were obtained:

Table 8.8 - GA results for transfer orbit parameters for non-planar elliptical initial and final orbits. (GA population size=150, number of generations=15, probability of crossover=0.9, probability of mutation=0.001.)

	GA Results
Eccentricity	0.2717
α_1	4.2387°
Semi-major axis	8835.6 km
$\Delta V_{\square\square\square}$	5.736 km/sec

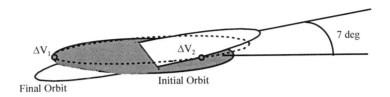

----- Transfer Orbit

Figure 8.19 – Initial and final orbits along with GA-produced transfer orbit.

Case III

Table 8.9 - Initial and final orbits parameters for two non-coplanar elliptical orbits.

	Initial Orbit	Final Orbit
Eccentricity	0.2	0.2
Inclination	5°	5°
Semi-major axis	8000 km	8500 km

When the GA is run on this case, it produced the following results (Table 8.10) (Figure 8.20):

Table 8.10 - GA results for transfer orbit parameters for non-planar elliptical initial and final orbits. (GA population size=150, number of generations=10, probability of crossover=0.9, probability of mutation=0.01.)

	Analytical Results	GA Results
Eccentricity	0.2289	0.241
α_1	0°	0°
Semi-major axis	8300 km	8432 km
$\Delta V_{\square\square\square}$	0.5963 km/sec	0.6643 km/sec

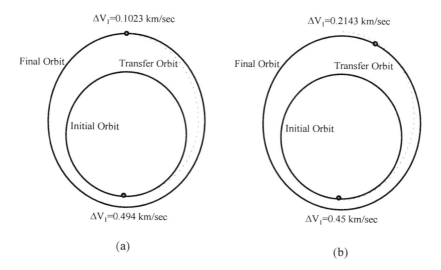

Figure 8.20 – (a) Initial and final orbits along with analytical transfer orbit (b) Initial and final orbits along with GA produced transfer orbit.

CONCLUSIONS

A genetic algorithm was used to search for transfer orbits which minimize the velocity change needed to transverse from one orbit to another. The effectiveness of using this approach was tested through comparison to two problems with known solutions. The GA produced near optimum results for the coplanar and non-coplanar Hohmann transfer problems. The GA was then used to search for transfer orbits between non-circular orbits which are inclined. While no analytical solutions exist for such problems, the resulting GA solution appeared quite reasonable. Therefore, it can be concluded that a genetic algorithm can be used successfully to find near optimum transfer orbits for ΔV_{TOT} requirements.

REFERENCES

[1] Chobotov, V. A. (1991). *Orbital mechanics*, Washington, DC: American Institute of Aeronautics and Astronautics, Inc.

[2] Bender, D.F. (1962). Optimum coplanar two-impulse transfers between elliptic orbits. *Aerospace Engineering*, **21**(Oct.), 44-52.

[3] Lawden, D. S. (1962). Impulsive transfer between elliptical orbits. *Optimization Techniques* (Chapter 11), edited by G. Leitmann, New York: Academic Press.

[4] Baker, J. M. (1966). Orbit transfer and rendezvous maneuvers between inclined circular orbits. *Journal of Spacecraft and Rockets*, **3**, 1216-1220.

[5] Reichert, A. (1993). Optimum two-impulse transfer between coplanar, non-aligned elliptical orbits. Paper presented at the 1993 Southeast Regional AIAA Conference, Tuscaloosa, AL.

[6] Pinon, E. III, & Fowler, W. T. (1995). Use of a genetic algorithm to generate earth to moon trajectories. *AAS Paper Number 95-141*.

[7] Bate, R. R., Mueller, D. D., & White, J. E. (1971). *Fundamentals of astrodynamics*. New York: Dover Publications Inc.

[8] Gerald, C. F., & Wheatley, P. O. (1989). *Applied numerical analysis, 4th Edition* (pp. 141-145). Reading, MA: Addison-Wesley Publishing Company.

[9] Goldberg, D. E. (1989). *Genetic algorithms in search, optimization, and machine learning*. Reading, MA: Addison-Wesley Publishing Company.

CHAPTER 9

DATA MINING USING GENETIC ALGORITHMS

Darrin J. Marshall

NCR Corporation
17095 Via Del Campo
San Diego, CA 92127
email: darrin.marshall@sandiegoca.ncr.com

ABSTRACT

Data mining involves sifting through very large databases in search of frequently occurring patterns to detect trends and produce generalizations about the data content. This chapter describes an approach, using genetic algorithms, to search through, or "mine," a large set of database transactions in order to determine a relationship between the transactions. Consider an example database consisting of transactions representing products purchased at a retail store. This genetic algorithm implementation focuses on determining, out of 100 possible items, the four items that are most often purchased together. Information of this sort can be used by businesses to aid in inventory, sales, and marketing strategy planning. For example, such information can allow a business to focus on marketing opportunities that encompass sets or groups of products and thus maximize their advertising dollars.

INTRODUCTION

With the widespread proliferation of powerful and affordable computing and information gathering devices, we have seen a dramatic increase in the amount of electronic data that is being obtained and stored. It has been estimated that the amount of electronic information in the world doubles every 20 months, and the size and number of databases that store this information are increasing even faster [1].

Given that we have vast amounts of data stored and available to us, and new mountains of information are being gathered each day with every credit card and point-of-sale bar code transaction, businesses are now faced with the task of making use of all this information. For instance, one of the countries largest retailers, Wal-Mart, as of 1995, had 65 weeks of point-of-sale transaction data on-line, taking up more than 3.5 terabytes of storage [2]. Considering that one terabyte of data is equivalent to approximately two million books, it is obvious that new and powerful techniques will be required to make effective use of this much information.

It is generally recognized within the business community that there is untapped value in large databases [3]. It is also recognized that this information is vital to business operations, and that decision-makers need to make use of the stored data in order to be competitive. To accomplish this, the large businesses that own these terabytes of data have turned to what is being called "data mining." Data mining employs methodologies to sift through huge databases in search of frequently occurring patterns to detect trends and produce generalizations about the data content. As one might suspect, this type of analysis not only requires the use of very sophisticated and powerful computers, designed specifically for the purpose of managing large amounts of data, but it also requires data mining application software that can efficiently and effectively search through vast amounts of data in order to recognize the possible relationships that exist.

Data mining encompasses many technologies and techniques which are used to identify "hidden" pieces of information within a database. Current technologies and techniques employed in many data mining environments include parallel computer architectures, relational database management systems (RDMS), data warehouses (which organize data in such a way that retrieval and analysis is greatly facilitated), and often, artificial intelligence and neural networks. These technologies, although important in the overall implementation of a data mining environment, are beyond the scope of the application examined in this chapter.

This chapter focuses solely on a genetic algorithm-based search technique that can be used in data mining to aid in determining relationships that may exist within a data set. In order to determine what relationships exist, many combinations of transaction comparisons and database queries must take place. Depending on the nature of the data and the relationships that are being sought, millions of data attribute combinations may have to be investigated.

For example, consider a database that contains N transactions, where each transaction consists of a list of items purchased. Also, assume that there are 100 different items that can appear in a transaction. Now, consider an effort to determine the four items that are most often purchased together. This implies that there are 100 items chosen 4 at a time, or 3,921,225 combinations of items available for investigation. If any useful output from the data mining tool is expected within a reasonable amount of time, an exhaustive comparison of all possible combinations is probably not feasible. Instead, an efficient and effective search technique is required for relationship discovery. This is where a genetic algorithm comes into play, and it is this problem that is addressed in this chapter.

The remainder of this chapter will review current data mining search techniques, describe the problem that was solved using a genetic algorithm, present the specifics of the genetic algorithm that was developed, and present the results of simulations that were run to test and analyze different aspects of the genetic algorithm in the data mining problem domain.

REVIEW OF CURRENT DATA MINING SEARCH TECHNIQUES

In order to better appreciate the importance and relevance of the genetic algorithm search approach proposed in this chapter, and to better understand the aspect of search techniques and how they relate to data mining, several issues related to the searching of data will be reviewed. Specifically, the importance of search in data mining and current approaches to search in data mining will be discussed.

Importance of Search in Data Mining

Data mining is a field still in its formative stages. Because of this, the actual meaning of the term "data mining" is open to interpretation. Some consider data mining to be a component within a bigger "knowledge discovery" process, while others consider data mining and knowledge discovery to be one in the same. In addition, because data mining is in its infancy, data mining techniques are still being developed and investigated. Nonetheless, however one chooses to define data mining, it is a fact that efficient and effective search mechanisms are an important and essential component in the process of discovering potential relationships that exist within large data sets.

As previously mentioned, the explosive growth in our ability to collect electronic information has far outpaced our ability to interpret and make use of that information. This, in turn, has created a need for a new generation of tools and techniques for database analysis. One such technique, data mining, employs methodologies to sift through huge databases in search of frequently occurring patterns to detect trends and produce generalizations about the data content.

As one might suspect, in order to find patterns, detect trends, or produce generalizations about data content, extensive searching, not only of the actual data, but of various aspects of the data, must take place. In the case of data mining, search is the process of seeking a solution by examining alternatives. In the case of the problem investigated in this chapter, many item combination alternatives must be examined in order to find a combination that meets the given constraint. In the case of our example, the constraint is to determine the four most frequently occurring items that appear within the transactions of a database. Because of the huge number of combinations and alternatives that typically need to be investigated in data mining problem domains, the need for effective and efficient search mechanisms is of extreme importance. Therefore, since genetic algorithms are excellent search mechanisms, their application in data mining seems "natural."

Current Approaches to Search in Data Mining

When considering search mechanisms in a data mining environment, a distinction must be made between different types of search. For the purposes of this discussion, a distinction will be made between a "low level" search of the database (which is not the focus of this chapter) and a "high level" search of alternatives.

In a data mining environment, there are different levels of search involved in the extraction of information from a database. At the lowest level, when specific files or records need to be accessed, access structures called indexes, are used to speed up the retrieval of records in response to certain search conditions. The idea behind an index access structure is similar to that behind the indexes used in textbooks. The index contains a key term along with a page number, or list of page numbers where the key term can be found. We can search the index to find an address (page number in a textbook) or list of addresses and then use the address(es) to locate the term in the database (the textbook). The (unacceptable) alternative is to search the whole database (the textbook) to find the term we are interested in [4].

Since (as discussed later in the section, "Problem Statement") an actual database is not being used for purposes of this project, the lowest level of database search becomes trivial and the indexing concept described above is not necessary. In a "real" database environment, however, indexing mechanisms would be present.

A "high level" search, on the other hand, is a search mechanism that operates above the lower level searching mechanism just described. A high level search, rather than performing the actual search through a database, determines *what* the lower level search should look for. The focus of this discussion is on the "high level" search aspect of a data mining environment, and it is the genetic algorithm that can be used for such a search. Figure 9.1 shows how these search levels fit into the overall data mining environment.

Data Mining Environment

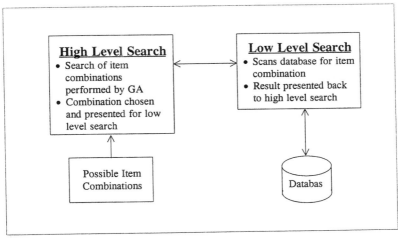

Figure 9.1 - Search levels in the data mining environment.

There are many possible approaches for implementing high level search in a data mining environment. The following are some definitions used in the discussion of high level search strategies. Following the definitions are several high level search strategies that have been cited as popular alternatives for implementation in present day data mining environments [5]:

Definitions:

- **Search space:**   A search space consists of a set of states (nodes) and a set of moves to neighboring states. Search spaces may consist of pre-existing data structures that represent states which can be visited during the search, and moves which guide the traversal of the search space.

- **Search strategy:**   A search strategy is a general approach to construct and process a search space. For example, a heuristic search strategy differs from an exhaustive search strategy by leaving part of the search space unsearched.

- **Heuristic search:**   A heuristic search generates and/or processes only part of a total search space. Hill climbing and Beam search, each described below, are considered heuristic search approaches. Given this definition, a genetic algorithm would be considered a heuristic search.

- **Exhaustive (or Enumerative) search:** An exhaustive search is a search that can reach each node of a search space, ensuring the optimal solution. This approach is usually not feasible when the search space is extremely large, as is typically the case in data mining applications.

High Level Search Strategies:

- **Hill climbing:** Hill climbing is a heuristic search strategy which retains one node after each step of search. At each step, it selects the best neighbor of the current state, according to an optimizing criterion. If each neighbor is worse than the current state, the search halts.

- **Beam search:** Beam search is a heuristic search strategy similar to hill climbing. At each step, the best N partial solutions are retained for further search according to an optimizing criterion.

- **Exhaustive (or Enumerative) search:** (described above)

- **Hierarchical tree search:** A tree partitions a data set into a hierarchically ordered set of possible alternatives where each alternative on a hierarchical level is recursively divided into sub-concepts on the next lower hierarchical level. A hierarchical tree search algorithm, "HierarchyScan," proposes an alternative to a traditional sequential scan that is typically used to search a tree [6].

PROBLEM STATEMENT

The problem addressed in this chapter consists of using a genetic algorithm to search a synthetically generated database in order to discover a relationship between data items. The synthetic database contains N transactions, where N varies based on the test being run. Each transaction, similar to transactions found in a retailer's database, consists of a list of items purchased, where each item is one of 100 different items. Given a database with the characteristics just described, the goal is then to determine the four items that are most often purchased together.

The remainder of this section discusses an overview of the problem implementation, the format of a synthetic data set, and a supporting program, the synthetic data generator.

Implementation Overview

If we are to find four related items out of a possible 100, this implies that there are 100 choose 4, or 3,921,225, combinations of items that we can potentially investigate. As previously mentioned, it is unreasonable to investigate all 3,921,225 combinations of potential relationships since, for each combination, the entire database (or a sufficiently large subset) has to be scanned to determine how often the four items appear together. To address this problem, a genetic algorithm is used to investigate a subset of the possible combinations. The intent was that by using a genetic-based search, the combination of four items that are most often purchased together could be determined much more efficiently than through an enumerative investigation of every possible combination. Genetic algorithm specifics on coding, fitness, and operators is given in the following section, "Genetic Algorithm Specifics." Thus, an overview of the genetic algorithm in the data mining problem domain is presented next.

Before performing any type of genetic search, a coding scheme was determined to represent the four potentially related items. To prepare for the search, X sets (the population size) of four items are picked at random and then coded into strings based on the coding scheme. These X strings represent generation 0 of the search. Then, for each of the X sets of four coded items, the entire database (or a sufficiently large subset) is scanned to generate a statistic (the fitness) as to how often the four items are purchased together. This fitness is based on the frequency with which the set of four items appear within the database under investigation. In addition, if a fractional number of the items in the item list appear within a transaction, the fitness reflects that fraction.

Once a fitness value is determined for each set of items in the population, the three genetic operators (reproduction, crossover, and mutation) are performed to generate a new population of strings. After performing reproduction, crossover, and mutation, if all goes as planned, each successive population should contain more highly fit strings, ultimately leading to a set of the four most frequently occurring items within the database.

In order to fully verify and analyze this genetic algorithm approach, simulations were run using several different fitness functions, crossover operators, mutation operators and genetic algorithm parameters. In addition, simulations were run using several different size data sets with varying characteristics (see the database configuration matrix in the section, "Results" for details). The synthetically generated data, as described in the next section, provided the advantage of being able to represent specific relationships within a data set and facilitated verification of the genetic algorithm-based search approach.

Synthetic Data Sets and Data Set Generation

Since the focus of this chapter is on the development and verification of a genetic algorithm-based data mining search approach, an actual database with "real" data was not used nor was it required to demonstrate the usefulness of the genetic algorithm. Instead, "synthetic" data sets were generated to demonstrate the application of the genetic algorithm. As described earlier, a synthetically generated data set consists of X transactions. Each transaction consists of a transaction number (Trans_Num), the number of items purchased in the transaction (N), and a list of N items, where each item is specified by a number (i.e. 0=Pretzels, 1=Aspirin, 2=Beer, 3=Bread, etc.). Also, N can be any of 100 different items. The format of a transaction and an example of a transaction containing fifteen items are shown below:

Transaction Format: Trans_Num N $Item_0$ $Item_1$ ... $Item_N$
Example Transaction: 0 15 1 2 3 10 19 55 42 41 22 83 20 20 87 92 6

Synthetic data sets were generated using a C program that allowed the user, through command line input, to specify data characteristics. Allowing the user control over data set characteristics facilitated verification of the genetic algorithm search results, since known relationships were embedded within the data sets. Then, the relationships that were discovered by the genetic search were confirmed against the expected results.

The synthetic data generator program required the user to provide at least one option to specify the number of transactions to generate. Other options allowed the user to specify individual items and the probability with which each of the specified items appeared within the synthetically generated data set.

GENETIC ALGORITHM SPECIFICS

The coding scheme, fitness function, and each of the three genetic operators (reproduction, crossover, and mutation) can be implemented in a variety of ways depending on the problem to which a genetic algorithm is being applied. Various implementation alternatives of these for the data mining problem examined in this chapter are discussed below.

Parameter Coding

A main difference between genetic algorithms and more traditional optimization and search algorithms is that genetic algorithms work with a coding of the parameter set and not the parameters themselves. Thus, before any type of genetic search can be performed, a coding scheme must be determined to represent the parameters in the problem at hand. In the data mining problem addressed by this chapter, the parameters of interest are simply four, potentially related, item numbers. Therefore, a coding scheme for four item numbers was determined considering the following factors:

- A multi-parameter coding, consisting of four sub-strings, is required to code each of the four items into a single string.

- Each sub-string needs to represent one of a hundred (0 through 99) possible items.

- There are 100 choose 4 (3,921,225) possible combinations of items that need to be represented.

Given these factors, several coding schemes were considered. First, a scheme using a base-10 coded sub-string was examined. This coding allows 100 items to be represented, and no codings correspond to non-existent items (all possible sub-string codings would represent valid item numbers, 0 through 99). This coding results in an 8-digit string where each digit (d_x) is between 0 and 9, as follows:

$d_1 d_0$	$d_1 d_0$	$d_1 d_0$	$d_1 d_0$
item 1	item 2	item 3	item 4

A coding of this sort, however, limits the number of schemata that are available for the genetic algorithm to exploit. Therefore, since we want to maximize the number of schemata, a binary coding, as shown below was considered.

$b_6 b_5 b_4 b_3 b_2 b_1 b_0$	$b_6 b_5 b_4 b_3 b_2 b_1 b_0$	$b_6 b_5 b_4 b_3 b_2 b_1 b_0$	$b_6 b_5 b_4 b_3 b_2 b_1 b_0$
item 1	item 2	item 3	item 4

Here, each digit (b_x) is a 1 or 0. One problem with this coding is that, in order to represent 100 items, a sub-string must consist of seven bits ($2^7 = 128$). This means that 28^4 (128-100=28 values in each sub-string), or 614,656 strings could be manipulated by the genetic algorithm which don't correspond to actual solutions.

Another possibility, when considering the fact that there are 3,921,225 combinations of items that need to be represented, was to code the items as a 22-bit binary string. This corresponds to 2^{22}, or 4,194,304 possible solution representations, reducing the number of non-solution strings to 273,079 (4,194,304-3,921,225). Using this approach, however, requires a mapping from a "combination number" to the actual four items represented by that combination since the item numbers are not specifically embedded within the string. If this combination-mapping approach was used, it would be difficult for the genetic algorithm to exploit similarities between strings since two combination numbers would not necessarily have anything in common, even if the items represented by the combination numbers were similar. Therefore, this coding approach was deemed unacceptable.

Given these possibilities, and considering the implementation of the crossover operator which is discussed in the section, "Reproduction, Crossover, and Mutation Operators," the decimal coding appeared to be the best alternative and was therefore implemented for the data mining problem.

Fitness function

If a string of four items is coded as just described, the fitness for such a string can be determined based on the frequency with which the set of four items appear within the transactions contained in the database under investigation. In addition, if a fractional number of the items in the item list appear within a transaction, the fitness value will also reflect that fraction. For example, if the item list for which a fitness value is to be determined is as follows:

01	22	68	07

and if the database, for the purpose of this discussion, contains only two transactions, as follows:

transaction 1:   34 22 55 01 68 99 02 07 42 24
transaction 2:   76 01 86 99 02 07 42 24

then transaction 1 contains all four items (01, 22, 68, and 07) and transaction 2 contains only two of the four items (01 and 07). If $I(t_i)$ represents the fraction of items that are contained within a single transaction t_i, then $I(t_1)$ would equal 1.0 and $I(t_2)$ would equal 0.5 (2/4). If all the $I(t_i)$s are summed over all transactions, and this sum is divided by the total number of transaction, N, then a relationship (referred to as the *base fitness*) between the item set and the frequency with which the items appear within the database can be obtained. Thus, the base fitness value for the item combination of 01, 22, 68, and 07 for the example database of two transactions would be:

$$I(t_1) + I(t_2) / N \quad \Rightarrow \quad (1.0 + 0.5) / 2 = 0.75$$

This yields a base fitness function that can be represented as follows:

Assume:
N = total number of transactions in the database file (sdg.db)
t_i = transaction number
$I(t_i)$ = fraction of items that are contained in t_i
M = sum of all $I(t_i)$ over all transactions

then M is:

$$M = \sum_{i=1}^{N} I(t_i)$$

and the fitness, referred to as F1, for a given string of items is:

$$F1 = base\ fitness(item\ string) = \frac{M}{N}$$

Furthermore, in an effort to avoid possible premature convergence, and to promote competition between strings throughout a simulation, a scaled fitness, F2, was implemented and tested. F2 is expressed as,

$$F2 = a*F1 + b$$

where $F2$ is the scaled fitness, $F1$ is the raw fitness, and a and b are determined based on the maximum and average raw fitness values. In the case of the data mining problem, the maximum possible fitness (if every transaction in the database contained all four items of interest) was 1.0 and the average raw fitness was found to be around 0.4.

In a third alternative, the fitness function was based on the following additional assumptions: Since the goal is to determine the four items that are most often purchased together, the fitness function should yield a higher value if more of the items in the item set are contained together within the database transactions. For example, if one item set yielded a base fitness value of 0.25 (implying that, on average, one of the four items in the item set were contained in each of the database transactions) and a second item set yielded a base fitness value of 0.75 (implying that, on average, three of the four items in the item set were contained in each of the database transactions), it may be desirable to give more than a simple linear emphasis to the second item set value since it is closer to the goal of four. Thus, a third fitness, referred to as F3, is expressed as

$$F3 = \frac{M^x}{N}$$

where x was chosen to be 3.

While this type of fitness expression gives more emphasis to item sets that are closer to the four item goal, it must be realized that a single occurrence of four items purchased together within a large database will not yield a higher fitness than three of the items purchased together many times. Since we are seeking trends within a large set of database transactions, it was felt to be more important to recognize major trends (such as three items purchased together many times) which might lead to a trend of four items purchased together.

The results of simulations run with each of the fitness expressions just described (F1, F2, and F3) are discussed in the section, "Fitness Function Simulation Results." Those results show that fitness function F3 provided very good performance, better than both F1 and F2.

Reproduction, Crossover, and Mutation Operators

The reproduction, crossover, and mutation operators that were implemented to support the eight-digit decimal coding are discussed next.

Reproduction Operator

A basic roulette wheel reproduction operator was implemented for this data mining application. In this type of reproduction, each string in the population is given a roulette wheel slot sized in proportion to its fitness. Then, by "spinning" the wheel N times (the population size), N new offspring are created for the next generation. By having a weighted, or "biased" roulette wheel, it is more probable that higher fit strings receive more copies in subsequent generations.

Crossover Operator

Crossover is used to improve the population fitness by introducing new strings into the population. Thus, the crossover operator for this problem had to introduce new item combination strings into the population. Due to the fact that the sub-strings (item numbers) within a coded string must be preserved, possible cross sites exist at the item boundaries within a string, as shown by points A, B, and C below:

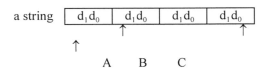

In determining a crossover operator for this application, it was important to consider the fact that duplicate item numbers cannot exist within a string since this would represent an invalid combination of items. For example, if the following two strings were crossed using simple crossover and a cross site as noted by A,

	04	21	06	15
string 1	04	21	06	15
string 2	03	93	24	04

A

strings 1' and 2' would result, as shown:

string 1'	04	21	24	04
string 2'	03	93	06	15

Note that the resulting string 1' only contains three different item numbers (04, 21, and 24) instead of four.

Given this consideration, it was obvious that a simple single point crossover operator would not be adequate. Crossover operators such as partially matched crossover, or PMX [7], and ordered crossover [7] ensure that duplicates do not occur in children strings. However, these crossover operators require that each parent string contain the same characters. The situation in the data mining problem, however, is slightly different in that every item in the first string may, or may not, be in the second string. Considering this fact, two different cross-over operators were developed for examination: aligned single-point crossover (ASPX) and unmatched crossover with single child offspring (UXSCO), each of which is described below. The results of simulations run with each of these crossover operators are discussed in the section, "Crossover Operator Simulation Results." Those results show that the ASPX crossover operator provided very good performance, better than the UXSCO operator.

Aligned Single-Point Crossover (ASPX)

Recall that the main goal of our crossover operator is that it produces children strings which do not contain duplicate item numbers. This goal can be achieved with aligned single point crossover as follows: Two parent strings, such as those shown below, are chosen at random from the current population.

string 1	10	20	30	40
string 2	88	10	20	99

Next, since the order of the item numbers within a string is not important, a string can be rearranged without affecting its fitness value. Thus, the item numbers in string 2 are rearranged such that each item number that has a corresponding match in string 1 is aligned with that item number. In this example, the 10 and 20 in string 2 are matched in string 1 and would thus be aligned, as follows:

string 1	10	20	30	40
string 2'	10	20	88	99

Now, since all matching item numbers are aligned between the two strings, basic single-point crossover can be performed and guarantees that no duplicates occur in either child string. For example, if a cross site is chosen at point A,

string 1	10	20	30	40
string 2'	10	20	88	99

the following children strings will result and no duplicates will be present.

string 1'	10	20	30	99
string 2''	10	20	88	40

Unmatched Crossover with Single Child Offspring (UXSCO)

As with aligned single-point crossover, unmatched crossover with single child offspring operates with the main goal of producing children strings which do not contain duplicate item numbers. In unmatched crossover with single child offspring, this is achieved as follows: Again, two parent strings are chosen, at random, from the current population. Next, it is determined if there are any matching items between the two strings. As part of this determination, the item numbers that are matched between the two strings are saved. In addition, the item numbers in string 2 that do not match any items in string 1 are saved. Next, a random number, X, between one and the number of items that are unmatched is generated. If there are zero unmatched items (the parent strings are identical) then X will be zero and no crossover will take place. If there are greater than zero unmatched items (the parent strings are not identical), then X unmatched item numbers from parent 2 are copied to random locations in child 1. Following the copy of X items to child 1, child 2 will be equal to parent 2 (there is a single new offspring, child 1). The following example demonstrates the operation of unmatched crossover with single child offspring:

Assume the following two strings, which possess two matched items (10 and 20) and two unmatched items, were chosen:

string 1	10	20	30	40
string 2	88	10	20	99

Next, assume that choosing a random number between 1 and the number of unmatched items (two) results in X equal to two. Then, two unmatched items from string 2 (88 and 99) are copied to two random locations in string 1. Assuming the random locations are one and three, the following children strings will result:

string 1'	88	20	99	40
string 2'	88	10	20	99

By performing crossover in this manner, it is guaranteed that no duplicate item numbers will appear in a string.

Crossover Operator Support of Coding Scheme

Both the ASPX and UXSCO operators supported the choice of a decimal coding over a binary coding. Since the goal of crossover is to create strings with new item number combinations, there would be no advantage to allowing crossover to occur at sites other than between item number boundaries. Given this, there would be no advantage to having a string coded in binary form since no additional "useful" schemata would be processed. Using a binary coding, more schemata would be available for processing by the genetic algorithm, however, the additional available schemata would represent strings that are not possible solutions to the problem.

Mutation Operator

With the use of a base-10 coding as previously described, an alternate method of mutation was required. In a typical binary coding, mutation is simple in that if it is determined that mutation should take place, a 1 is mutated to a 0, and a 0 is mutated to a 1. In a base-10 coding, however, a method to mutate a decimal value between 0 and 99 must be determined. The following alternatives were considered:

1) generate a random number between 0 and 99 ("random" mutation)

2) generate a new number within some $\pm\delta$ of the number being mutated ("window" mutation)

In an effort to determine if there was an advantage of one mutation operator over the other, each was examined through simulations and the results are discussed in the section, "Mutation Operator Simulation Results." For the most part, each mutation operator performed equally well.

RESULTS

The following section discusses the different aspects of the genetic algorithm that were investigated in an effort to select the best performing fitness function and genetic operators. In addition, the specific simulations that were run for each area of investigation are discussed and summarized.

Areas of Investigation and Simulations

In an effort to better understand and optimize the data mining genetic algorithm, several aspects of its implementation were investigated. Areas that were investigated included fitness function, crossover operator, mutation operator, genetic algorithm parameters, and genetic algorithm performance.

Various simulations were run using different combinations of genetic algorithm operators, genetic algorithm parameters, and database configurations. The database configuration table (Table 9.1) and simulation test matrix (Table 9.2) summarize the simulation configurations that were run. A discussion of each area of investigation, the specific simulations that were run for each area, and the results that were obtained is presented following the database configuration table and simulation test matrix.

Database Configurations

The following table summarizes the different database configurations that were used in the simulations specified in the simulation test matrix:

Table 9.1 - Database configuration table.

Database Configuration	Number of Transactions	Item Relationship(s)
1	100	Items 11, 22, 33, 44. 90% probability for each.
2	100	Items 11, 22, 33, 44. 60% probability for each.
3	500	Items 11, 22, 33, 44. 90% probability for each. Items 55, 66, 77, 88. 80% probability for each.
4	5000	Items 11, 22, 33, 44. 60% probability for each.

Each database configuration contains a different type of item relationship where each relationship attempts to "challenge," in a different way, the genetic algorithm in determining the four items that most often appear together. For example, database configuration 1 was generated such that each of the items 11, 22, 33, and 44 appear in a single transaction with a 90% probability. This implies that the probability of all four items appearing together in a single transaction is 0.9 x 0.9 x 0.9 x 0.9, or 0.656 (65.6%). This configuration is intended to be the simplest database from which the four items could be determined. The item relationships for the remaining database configurations can be determined in a similar manner using the information provided in the database configuration table.

Simulation Test Matrix

The following simulation test matrix (Table 9.2) shows the simulations that were run for each area of investigation (fitness function, crossover operator, mutation operator, genetic algorithm parameters, and genetic algorithm performance). The test case numbers are referred to in the subsequent sections that discuss the results of each simulation.

Table 9.2 - Simulation test matrix.

Fitness Function Simulations

Test Case	Database Characteristics	GA Variables	Fitness Function	Crossover Operator	Mutation Operator
1	Database configuration 1	Pop. Size: 70, p_c : 0.9, p_m: 0.001	F1	UXSCO	Random
2	Database configuration 2	Pop. Size: 70, p_c : 0.9, p_m: 0.001	F1	UXSCO	Random
3	Database configuration 3	Pop. Size: 70, p_c : 0.9, p_m: 0.001	F1	UXSCO	Random
4	Database configuration 1	Pop. Size: 70, p_c : 0.9, p_m: 0.001	F2	UXSCO	Random
5	Database configuration 2	Pop. Size: 70, p_c : 0.9, p_m: 0.001	F2	UXSCO	Random
6	Database configuration 3	Pop. Size: 70, p_c : 0.9, p_m: 0.001	F2	UXSCO	Random
7	Database configuration 1	Pop. Size: 70, p_c : 0.9, p_m: 0.001	F3	UXSCO	Random
8	Database configuration 2	Pop. Size: 70, p_c : 0.9, p_m: 0.001	F3	UXSCO	Random
9	Database configuration 3	Pop. Size: 70, p_c : 0.9, p_m: 0.001	F3	UXSCO	Random

Crossover Operator Simulations

Test Case	Database Characteristics	GA Variables	Fitness Function	Crossover Operator	Mutation Operator
7, 8, 9	*Note: Same as Fitness Function Simulations 7, 8, and 9*				
10	Database configuration 1	Pop. Size: 70, p_c : 0.9, p_m: 0.001	F3	ASPX	Random
11	Database configuration 2	Pop. Size: 70, p_c : 0.9, p_m: 0.001	F3	ASPX	Random
12	Database configuration 3	Pop. Size: 70, p_c : 0.9, p_m: 0.001	F3	ASPX	Random

Mutation Operator Simulations

Test Case	Database Characteris- tics	GA Variables	Fitness Function	Crossover Operator	Mutation Operator
11	Database configura- tion 2	Pop. Size: 70, p_c : 0.9, p_m: 0.001	F3	ASPX	Random
13	Database configura- tion 2	Pop. Size: 70, p_c : 0.9, p_m: 0.001	F3	ASPX	Window

Genetic Algorithm Parameters Simulations

Test Case	Database Characteris- tics	GA Variables	Fitness Function	Crossover Operator	Mutation Operator
14	Database configura- tion 2	Pop. Size: 200, p_c : 0.9, p_m: 0.001	F3	ASPX	Random
15	Database configura- tion 2	Pop. Size: 70, p_c : 0.9, p_m: 0.010	F3	ASPX	Random
16	Database configura- tion 2	Pop. Size: 70, p_c : 0.9, p_m: 0.010	F3	ASPX	Window

Performance Simulations

Test Case	Database Characteris- tics	GA Variables	Fitness Function	Crossover Operator	Mutation Operator
17	Database configura- tion 4	Pop. Size: 70, p_c : 0.9, p_m: 0.001	F3	ASPX	Random
18	Database configura- tion 4 *Note: Only a fraction (0.1) was sampled*	Pop. Size: 70, p_c : 0.9, p_m: 0.001	F3	ASPX	Random

Simulation Graph Legends

For each of the simulation graphs results that are shown in the following sections, the following legend applies, representing the average fitness, best fitness and the solution for each run:

Fitness Function Simulation Results

Three different fitness functions (F1, F2, and F3), as described previously in the section, "Fitness function," were investigated. The results of each fitness function simulation are discussed below.

F1 Fitness Function Simulations

Each F1 fitness function simulation was run with the crossover, mutation, and genetic algorithm parameters as noted in the simulation test matrix.

Simulation 1 was run with database configuration 1, which is a database consisting of 100 transactions and an embedded relationship between items 11, 22, 33, 44. Of the 100 transactions, 51 contained all four of the items. As can been seen from Figure 9.2, the item combination of interest entered the population in less than fifteen generations, and the entire population average temporarily converged on the solution around generation 160.

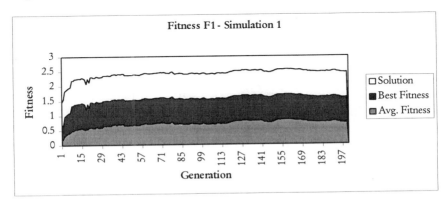

Figure 9.2 - Simulation 1 results

Simulation 2 was run with database configuration 2, which is a database consisting of 100 transactions and an embedded relationship between items 11, 22, 33, 44. Of the 100 transactions, 11 contained all four of the items. Thus, this database represented more of a challenge to the genetic algorithm than that of database configuration 1 since there were fewer occurrences of the four item combination. As can be seen from Figure 9.3, the simulation prematurely converged on an incorrect solution and did not locate the item combination of interest.

Fitness F1 - Simulation 2

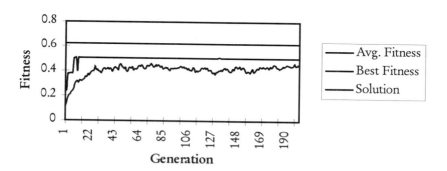

Figure 9.3 - Simulation 2 results

Simulation 3 was run with database configuration 3, which is a database consisting of 500 transactions and two embedded relationships; the first between items 11, 22, 33, 44 and the second between items 55, 66, 77, and 88. Of the 500 transactions, 290 contained all four of the items 11, 22, 33, and 44, and 85 contained all four of the items 55, 66, 77, and 88. Thus, this database, while it has many of the first item combination occurrences to guide the simulation, there are also many of the second item combination occurrences to provide a "distraction" to the genetic algorithm. As can be seen from Figure 9.4, the optimum solution (item combination 11, 22, 33, 44) was not located. The simulation, instead, prematurely converged on a solution consisting of members from each of the embedded item combinations (11, 33, 44, 88).

Fitness F1 - Simulation 3

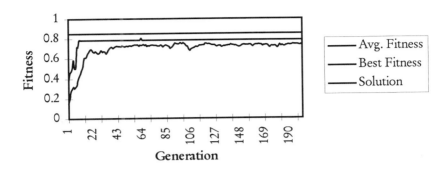

Figure 9.4 - Simulation 3 results

F1 Fitness Function Simulation Conclusions

Simulation 1, the easiest of the database configurations, converged on the correct solution but only temporarily. Simulations 2 and 3, on the other hand, prematurely converged on an incorrect solution. Therefore, in an effort to resolve this premature convergence, a second fitness function (F2) involving fitness scaling was implemented and is discussed in the next section.

F2 Fitness Function Simulations

Each F2 fitness function simulation was run with the crossover, mutation, and genetic algorithm parameters as noted in the simulation test matrix.

Simulation 4 was run with database configuration 1 (previously described). As can be seen from Figure 9.5, the simulation average increased much more gradually than in the corresponding unscaled simulation (simulation 1) which suggests that the scaling mechanism was preventing premature convergence, as intended. The item combination of interest entered the population around generation 110 but the simulation did not completely converge on the solution.

Fitness F2 - Simulation 4

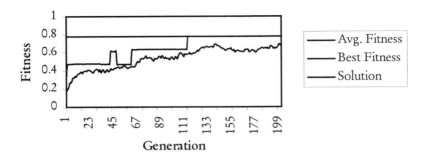

Figure 9.5 - Simulation 4 results

Simulation 5 was run with database configuration 2 (previously described). As can be seen from Figure 9.6, the simulation average did not converge as quickly as it had in the corresponding unscaled simulation (simulation 2) which suggests that the scaling was operating, as intended, to a certain degree. However, the simulation prematurely converged on an incorrect solution.

Fitness F2 - Simulation 5

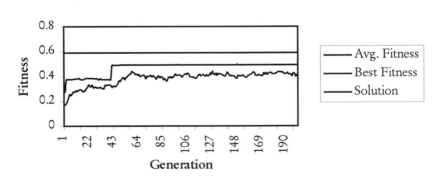

Figure 9.6 - Simulation 5 results

Simulation 6 was run with database configuration 3 (previously described). As can be seen from Figure 9.7, the item combination of interest never entered the population and the average fitness remained fairly constant throughout the simulation.

Fitness F2 - Simulation 6

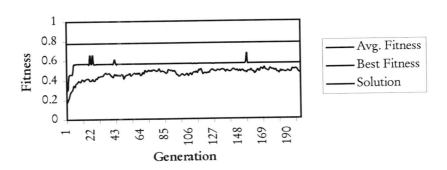

Figure 9.7 - Simulation 6 results

F2 Fitness Function Simulation Conclusions

None of the F2 fitness function simulations performed better than the corresponding F1 (unscaled) simulations. Thus, the fitness scaling mechanism did not improve simulation results as anticipated.

F3 Fitness Function Simulations

Each F3 fitness function simulation was run with the crossover, mutation, and genetic algorithm parameters as noted in the simulation test matrix.

Simulation 7 was run with database configuration 1 (previously described). As can be seen from Figure 9.8, the item combination of interest entered the population around generation 7 and the simulation quickly converged on the solution around generation 17.

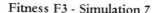

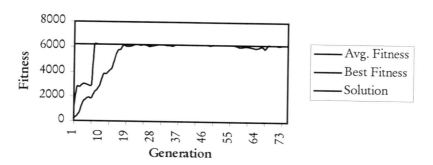

Figure 9.8 - Simulation 7 results

Simulation 8 was run with database configuration 2 (previously described). As can be seen from Figure 9.9, the item combination of interest entered the population around generation 5 and the simulation quickly converged on the solution around generation 20.

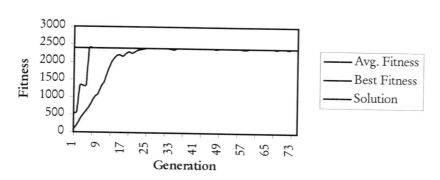

Figure 9.9 - Simulation 8 results

Simulation 9 was run with database configuration 3 (previously described). As can be seen from Figure 9.10, the item combination of interest entered the population around generation 15. While the population did not completely converge on the solution, there were a majority of solution strings within the population such that I would consider the solution to have been determined.

Fitness F3 - Simulation 9

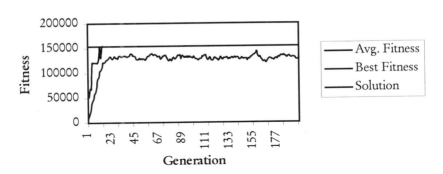

Figure 9.10 - Simulation 9 results

F3 Fitness Function Simulation Conclusions

The first two F3 fitness function simulations (simulations 7 and 8) converged on the correct solution while the third (simulation 9) did not completely converge but did perform well. Overall, fitness function F3 performed better than both fitness functions F1 and F2.

Fitness Function Simulation Summary

Of the three fitness functions that were investigated (F1, F2, and F3), fitness function F3 outperformed both F1 and F2. Thus, F3 appeared to be the best alternative and was therefore used in the remaining simulations.

Crossover Operator Simulation Results

As discussed in the prior section on crossover, two unique crossover operators, aligned single-point crossover (ASPX) and unmatched crossover with single child offspring (UXSCO), were developed and investigated. The results of each crossover operator simulation are discussed below.

UXSCO Simulations

Each of the preceding fitness function simulations (simulations 1 through 9) were performed with the UXSCO crossover operator. Thus, those simulation results will be used for comparison with the ASPX simulations and will not be duplicated here.

ASPX Simulations

Each ASPX crossover simulation was run with the fitness function, mutation, and genetic algorithm parameters as noted in the simulation test matrix.

Simulation 10 was run with database configuration 1. As can be seen from Figure 9.11, the item combination of interest entered the population around generation 5 and the simulation converged on the solution around generation 15. Comparing this to the corresponding UXSCO simulation (simulation 7), it can be seen that each simulation converged at essentially the same generation. In the ASPX simulation, however, it can be seen that the average fitness curve is much smoother than that of the UXSCO simulation. This is most likely due to the fact that the UXSCO operator is more random in nature than the ASPX operator.

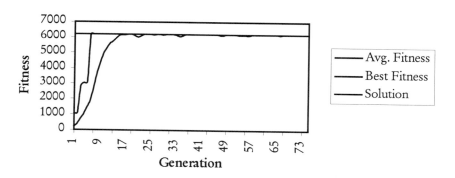

ASPX Crossover - Simulation 10

Figure 9.11 - Simulation 10 results

Simulation 11 consisted of two runs, 11A and 11B, each with database configuration 2. The only difference between runs 11A and 11B is that a different random number generator seed was used for each. As can be seen from Figure 9.12, the simulation converged on a false solution rather early on and never recovered (Note: This simulation was run for 400 generations, yet no improvement took place. Thus, only the first 75 generations are shown.).

ASPX Crossover - Simulation 11A

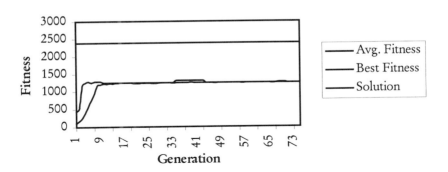

Figure 9.12 - Simulation 11A results

Simulation 11B, on the other hand, converged on the solution around genera-tion 12 (See Figure 9.13). Thus, it may have been the case that the initial population for run 11A was not diverse enough, and the mutation operator never introduced the lost allele into the population. This possibility is addressed in simulations 17 and 18 in the section, "Genetic Algorithm Parameter Simulation Results."

ASPX Crossover - Simulation 11B

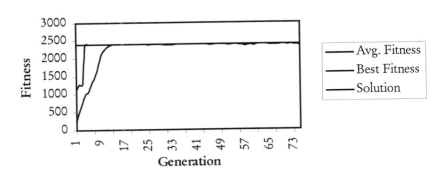

Figure 9.13 - Simulation 11B results

Comparing Simulation 11B to the corresponding UXSCO simulation (simula-tion 8), we can see that the ASPX simulation converged a bit sooner (generation 12 vs. generation 20).

Simulation 12 was run with database configuration 3. As can be seen from Figure 9.14, the item combination of interest entered the population around generation 8 and the simulation converged on the solution around generation 20. Comparing this to the corresponding UXSCO simulation (simulation 9), we can see that the ASPX simulation completely converged on the solution while in the UXSCO simulation, the population, even though it contained a majority of solution members, did not completely converge.

ASPX Crossover - Simulation 12

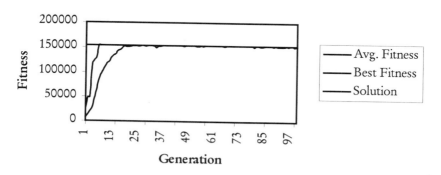

Figure 9.14 - Simulation 12 results

Crossover Operator Simulation Summary

Of the three simulation environments in which the ASPX and UXSCO crossover operators were tested, those implementing the ASPX operator performed as well or better than the corresponding UXSCO simulations. The first two simulations (using database configurations 1 and 2) converged on the solution around the same generation. In the simulations involving database configuration 3, however, the ASPX simulation clearly outperformed the UXSCO simulation in that the population totally converged on the solution. Therefore, due to the overall better performance of the ASPX crossover operator, it was used in the remaining simulations.

Mutation Operator Simulation Results

As discussed in the prior section on mutation, two unique mutation operators, random mutation and window mutation, were developed and investigated. The results of each mutation operator simulation are discussed below.

Random Mutation Simulations

Each of the previous simulations were run with the random mutation operator. Thus, those simulation results, namely the results from simulation 11A, will be used for comparison with the window mutation operator simulations and will not be duplicated here.

Window Mutation Simulations

The window mutation simulation was run with the fitness function, crossover operator, and genetic algorithm parameters as noted in the simulation test matrix.

Using the same test environment as that in simulation 11A, simulation 13 implemented the window mutation operator instead of the random mutation operator in an attempt to recover the allele that was assumed to be lost. Running with the window mutation operator, however, yielded essentially the same results as simulation 11A. In addition, the window mutation operator was run in several other simulation environments and in each case, the simulation results were essentially identical to those obtained in the corresponding random mutation operator simulations. Since the results were the same in each case, no graphs are shown.

Mutation Operator Simulation Summary

As suspected, no obvious advantage of one mutation operator over the other was found. Each corresponding random mutation and window mutation simulation converged (or did not converge) on the solution within the same number of generations. Thus, due to the equal performance of these two operators, the random mutation operator was used in the remaining simulations due to its simplicity.

Note: Mutation is further addressed, with respect to mutation probability, in the next section, "Genetic Algorithm Parameters Simulation Results."

Genetic Algorithm Parameters Simulation Results

Several combinations of genetic algorithm parameters were simulated, as specified in the simulation test matrix. Using the same database configuration and random number generator seed as in simulation 11A, two genetic algorithm parameters (population size and mutation probability) were modified in an attempt to see if those simulation results could be improved.

Population Size Simulations

Recall that in simulation 11A, the genetic algorithm converged on a false plateau and never recovered. Postulating that this could have been caused by an initial population that was not diverse enough, I ran a simulation in which the population size was increased from 70 to 200. As can be seen from Figure 9.15, the item combination of interest entered the population around generation 4 and the simulation converged on the solution around generation 13. Thus, the fact that this simulation converged on the solution suggests that a larger population may help insure that a correct solution is determined. Of course, it must be considered that a larger population has the disadvantage of contributing to more computation and thus a reduced overall execution time.

Population Size 200 - Simulation 14

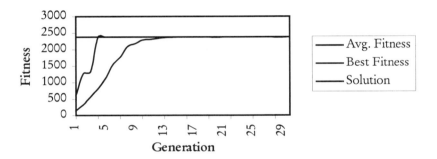

Figure 9.15 - Simulation 14 results

Mutation Probability Simulations (Random Mutation)

Again using simulation 11A as the test configuration, and using the random mutation operator, I attempted to recover the lost allele by increasing the mutation probability from 0.001 to 0.010. As can be seen from Figure 9.16, the item combination of interest entered the population around generation 190. Then the simulation, after having been stuck on a false plateau for most of the preceding simulation, quickly converged on the solution around generation 200. Thus, the fact that the simulation eventually converged on the solution suggests that an increased mutation probability may help insure that lost alleles are recovered and that a correct solution is determined.

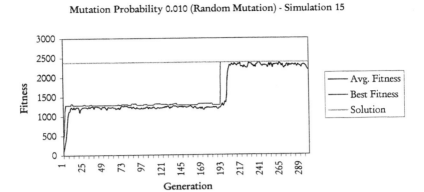

Figure 9.16 - Simulation 15 results

Mutation Probability Simulations (Window Mutation)

This simulation was identical to the previous simulation (simulation 15) except that the window mutation operator was used instead of the random mutation operator. As can be seen from Figure 9.17, the lost allele was never recovered and the simulation remained on a false plateau through 400 generations.

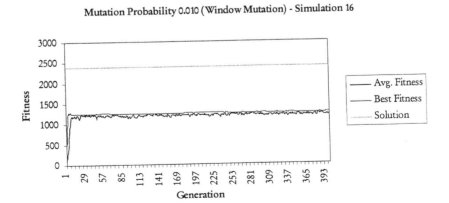

Figure 9.17 - Simulation 16 results

Genetic Algorithm Parameters Simulation Summary

In the environment in which the population size and mutation probability parameters were tested, it appears that increasing the population size can greatly increase the chances of converging on the correct solution, but at the cost of more computation. In addition, an increased mutation probability may help recover lost alleles and enable the simulation to locate the item combination of interest, as it did in the random mutation operator case.

Genetic Algorithm Performance and Performance Simulation Results

Given the fact that the data sets used in this project were synthetic, and also considering the limited scope of this project, the performance of the genetic algorithm-based approach could not be analyzed in a "real" database environment. One performance-related area that was investigated, however, involved improving the genetic algorithm execution speed. As is the case in most genetic algorithms, the bulk of the execution time is spent performing fitness function calculations, and the situation was no different for the genetic-based search examined in this project.

Depending on the size of the database that is being investigated by the genetic algorithm, the time to scan the database in order to generate the fitness for a particular item combination can vary significantly. If a relatively *small* database is being used, all transactions can be examined and an exact fitness value can be determined within a reasonable amount of time[1]. On the other hand, if the database is *very large*, complete examination of every transaction in order to calculate an exact fitness value, may not be feasible. In this case, we can perform an *approximate function evaluation* to help speed up the execution time of the genetic algorithm. In the case of the data mining genetic algorithm, approximate function evaluation is fairly straightforward in that a subset of the transactions in the database can be sampled to determine a fitness value for an item combination.

An approximate function evaluation method was implemented and compared to the full function evaluation method, and the simulation results that were obtained are discussed next.

---

[1] Of course, database size and fitness calculation times are relative, and the computer system on which a simulation is run will greatly impact overall execution time.

Full Function Evaluation Simulation

The full function evaluation simulation was run with the fitness function, crossover operator, mutation operator, genetic algorithm parameters, and database configuration as noted in the simulation test matrix.

Due to the large number of transactions in the database (5000), the total running time for this simulation was much longer than all previous simulations, at 43.5 minutes.

As can be seen from Figure 9.18, the item combination of interest entered the population around generation 5 and the simulation converged on the solution around generation 14.

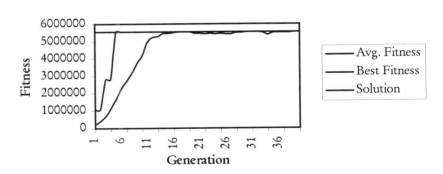

Full Function Evaluation - Simulation 17

Figure 9.18 - Simulation 17 results.

Approximate Function Evaluation Simulation

The approximate function evaluation simulation was run with the fitness function, crossover operator, mutation operator, genetic algorithm parameters, and database configuration as noted in the simulation test matrix.

The approximate function evaluation simulation used the same database configuration as that in the full function simulation; however, only one tenth (500) of the database transactions were sampled to generate the fitness for a given item combination. The total running time for this simulation was 5.1 minutes.

As can be seen from Figure 9.19, the item combination of interest entered the population around generation 7 and the simulation converged on the solution around generation 14. Comparing this to the corresponding simulation run with the full function evaluation, we can see that the approximate function evaluation simulation performed equally as well as the full function evaluation simulation in terms of convergence to the solution. In terms of execution speed, the approximate function evaluation simulation performed 8.53 times better than the full function evaluation simulation.

Approximate Function Evaluation - Simulation 18

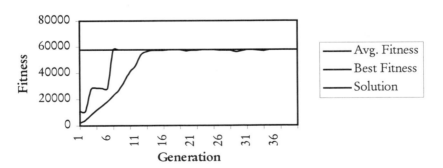

Figure 9.19 - Simulation 18 results.

Performance Simulation Summary

The approximate function evaluation simulation, for the data set examined in this test, converged on the solution in the same number of generations as the full function evaluation simulation. In terms of execution speed, however, the approximate function evaluation simulation performed 8.53 times better than the full function evaluation simulation. Thus, it appears that if the data set is fairly large, an approximate function evaluation method can find a solution much more efficiently than when a full function evaluation method is used.

SUMMARY

This chapter presented the results of a genetic algorithm implementation for a data mining application. The goal of this genetic algorithm implementation was to be able to determine, from a synthetically generated database, the four items that most often appeared together in a transaction. Through experimentation with a variety of fitness functions, crossover operators, mutation operators, and genetic algorithm parameters, the simulation results just presented provide evidence that this goal was achieved.

The development of the data mining genetic algorithm involved, first, the determination of a base-10 coding scheme to represent four potentially related items numbers. Then, in an effort to better understand the operation of the genetic algorithm and to identify the best fitness and genetic operators, multiple fitness functions (F1, F2, and F3), crossover operators (ASPX and UXSCO), and mutation operators (random and window) were developed and targeted for investigation through simulation. In addition to fitness and genetic operators, genetic algorithm parameters and performance were also targeted for investigation.

Also, a synthetic data generation tool was developed to enable production of data sets containing embedded relationships between specified item numbers. This capability greatly facilitated verification and analysis of the genetic algorithm implementation and simulations.

After determining the areas that were to be investigated, a simulation test matrix was created to define the tests and environments that were to be run. The execution of the defined tests led to the determination of a specific fitness function and set of genetic operators that provided the best results. The final fitness function and set of genetic operators that exhibited the best overall performance and determination of correct solutions included the following (each of which have been previously described):

Fitness function:	F3
Reproduction Operator:	Roulette Wheel
Crossover Operator:	Aligned Single-Point Crossover (ASPX)
Mutation Operator:	Random

Genetic Algorithm Parameters[2]

Population Size:	70 (200)
Crossover Probability:	0.9
Mutation Probability:	0.001 (0.010)

From the results achieved in this project, it is apparent that the use of genetic algorithms in data mining applications shows considerable promise and potential for further research and application. The genetic algorithm displayed its ability to quickly and efficiently locate and converge on correct solutions in a minimal number of generations. Future research might involve the implementation and analysis of this genetic algorithm approach in a "real" database environment. In addition, this project has further emphasized the fact that genetic algorithms are useful across a wide range of applications.

REFERENCES

[1] Dilly, R. (1996, February). *Data mining - An introduction.* The Queens University of Belfast web site
http://www-pcc.qub.ac.uk/tec/courses/datamining/ohp/dm-OHP-final_1.html.

---

[2] These genetic algorithm parameters provided the best "overall" performance. A higher population size (200) and mutation probability (0.010) did provide better performance in specific test scenarios.

[2] Hedberg, S. R. (Winter 1995). *Parallelism speeds data mining.* IEEE Parallel & Distributed Technology.

[3] Author unknown (1997). *Data mining - An IBM overview.* IBM web site http://www.almaden.ibm.com/cs.

[4] Elmasri. R., & Navathe, S. (1994). *Fundamentals of database systems.* Benjamin Cummings Publishers.

[5] Fayyad, U., Piatetsky-Shapiro, G., Smyth, P., & Uthurusamy, R. (1996). *Advances in knowledge discovery and data mining.* AAAI Press. The MIT Press.

[6] Chung-Sheng Li, Yu, P. S., & Castelli, V. (1996, 26 February through 1 March). HierarchyScan: A hierarchical similarity search algorithm for databases of long sequences. *Twelfth International Conference on Data Engineering.* New Orleans, LA, USA.

[7] Goldberg, D. E. (1989). *Genetic algorithms in search, optimization, and machine learning.* Reading, MA: Addison Wesley.

[8] Potter, W., Pitts, R., Gillis, P., Young, J., & Caramadre, J. (1992). *IDA-NET: An intelligent decision aid for battlefield communications network configuration.* New York: IEEE.

CHAPTER 10

SPACE SHUTTLE MAIN ENGINE CONDITION MONITORING USING GENETIC ALGORITHMS AND RADIAL BASIS FUNCTION NEURAL NETWORK

Daniel A. Benzing
Department of Aerospace Engineering and Mechanics
The University of Alabama
Tuscaloosa, Alabama

ABSTRACT

The Radial Basis Function Neural Network (RBFNN) has been utilized, with varying degrees of success, as a function approximator in many engineering applications. The principle discriminate which ultimately determines an RBFNN's viability for a particular case is its ability to construct higher order surfaces that can sufficiently perform the desired functional mapping. This can be difficult if the underlying relationship between the input and output variables exhibits an extremely nonlinear propensity. The general architecture of the RBFNN offers many parameters for its construction. While numerous methods for the optimization of these parameters have been documented, most have attempted to establish bits and pieces of the RBF structure, leaving some of the parameters to pure random choice, because of the immense computational cost incurred by the optimization of all parts. The genetic algorithm (GA) supplies an attractive means for accomplishing such a global parameter optimization. The purpose of this chapter is to establish a GA procedure that allows all the neurons of the RBF architecture to evolve simultaneously for the creation of higher order response surfaces that are accurate predictors and have adequate generalization ability. The hybrid architecture thus developed will be applied to a neural health monitoring system for the Space Shuttle Main Engine (SSME). In this case, the GA/RBFNN component will be required to quantify the metallic species content of the SSME plume, using the plume's electromagnetic emission. The underlying function which maps a plume's spectral signature to its metallic state is very nonlinear, and it will provide an excellent measure of the GA/RBFNN's ability.

INTRODUCTION

There are a multitude of problems in physics and engineering which involve the classic statistics problem of estimating a function from a set of input-output pairings. This task has typically been managed by traditional nonparametric methods; that is, there is no prior knowledge of the true function being approximated, and the free parameters in the model have no physical meaning connected with the problem at hand.   There are, however, neural techniques which incorporate supervised learning schemes that are capable of effectively learning extremely nonlinear mappings. Supervised learning asserts that the function is learned from a set of training patterns in conjunction with feedback from a teacher or critic.   These training patterns are a sequence of input (independent variable) to output (dependent variable) pairings:

$$\left\{ \left(\vec{x}_i, \vec{y}_i \right) \right\}_{i=1}^{P}$$

For a given input, the teacher or critic awards or penalizes the approximator based on comparison of the predicted outputs and the known correct outputs. Two established neural methodologies have met with success in this arena, namely: 1) traditional backpropagation and 2) RBFNN Classifiers.   In general, these methodologies can be considered nonparametric models.   Traditional backpropagation networks have multiple layers of activation functions, and use costly nonlinear gradient search methods in the free parameter optimization. Conversely, RBFNN networks have a single layer which is accompanied by a training methodology involving only basic linear algebraic concepts. This produces the desired effect of detaching the burden of lengthy calculations and provides an attractive, computationally inexpensive means of function approximation.

As shown in [1], there are times when the performance of an RBFNN can be significantly affected by perturbations of its internal parameters, with their establishment ultimately determining an application's success.   For the RBFNN, the basic internal parameters are the neuron activation functions, neuron placement in the functional space, and the extent of each neuron's coverage.   The optimization of all these parameters at once involves forbidding computational costs, and the job is typically reduced to only local optimizations of a few parameters.   The overall task of global parameter optimization, however, is suitable for a GA application.

The focus of the discussion herein will be the extension of the seminal work submitted by Whithead and Choate [2].   The foregoing paper establishes a co-operative-competitive GA that allows an RBFNN to evolve both its neurons positions and widths.   The procedure was tested on the benchmark Mackey-Glass differential equation [3] with the results showing the evolution of RBFNNs which had a 50-70% lower prediction error than those obtained using traditional k-means clustering [4].   The efforts of this chapter will involve the optimization of not only a neuron's position and width, but also to allow the neuron to take on a different form of radial response function.   Moreover, the chosen radial function will be given more freedom and control over its area of coverage.   The optimization of all Radial Basis Function (RBF) neurons will proceed simultaneously throughout the genetic evolution.

BACKGROUND

As discussed in the introduction, the objective of most neural networks is to estimate a function $y(\vec{x})$ from a training set of representative input/output pairings. Qualitatively, the RBF network does this by forming *localized* "bumps" or response regions within the input space. The superposition of the these local response regions forms a response surface that is of an order higher than the dimension of the input vector and spans the space covered by the input training patterns.

By definition, a radial basis function is one which decreases (or increases) monotonically away from a central point, thereby giving it an inherent bump form. Classic kernel functions (or in the case of the RBFNN, "neurons") that exhibit this propensity are the Gaussian, Cauchy, and the Inverse Multiquadric. These forms can be written generally as [5,6]:

A. Cauchy function,

$$g(z) = \frac{1}{1 + z^2} \qquad (10.1)$$

B. Inverse Multiquadric,

$$g(z) = \frac{1}{\sqrt{1 + z^2}} \qquad (10.2)$$

C. Gaussian function,

$$g(z) = \exp(-z^2) \qquad (10.3)$$

The form of z determines the type of radial scaling, or equivalently, the extent of the region influenced by the RBF.

Figure 10.1 delineates the Gaussian response function. To reiterate, the RBFNN positions a collection of these RBFs (in this case, Gaussians) through-out the space covered by the input training patterns. The parameter μ_j specifies the location of the RBF within the input space (μ_j has the same dimension as the input vector $\vec{x}$) and the parameter $\Box_j$ determines the width of the local function. Thus, a given RBF will be centered at μ_j within the input space and have a "receptive field" which is proportional to $\Box_j$. Moreover, it will give a maximum response for input vectors, $\vec{x}$, which are nearest the RBF center, μ_j.

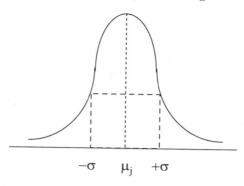

$-\sigma \quad \mu_j \quad +\sigma$

Figure 10.1 - Gaussian function.

By arranging an assortment of these receptive fields, response areas are created which sufficiently cover the input space; sufficient in the sense that the RBFNN can approximate the underlying function to within some pre-defined error criteria. More specifically, a complex decision hypersurface is constructed through the overlapping of the localized kernel regions. With a developed approximation surface, the RBFNN estimates an output, for an incoming input case, by first evaluating each of the kernel functions (in other words, determining where the input vector lies on the hypersurface) and then forming a weighted linear summation of their responses. The difficulty arises not from the logical evaluation of an input, but rather the establishment of the network parameters for the hypersurface construction, namely: center positions (μ_j), kernel widths (σ_i), and the weighting coefficients for the summation of the individual kernel responses.

The development of an RBFNN is done in a two-part learning scheme known as hybrid learning (Figure 10.2). The initial forward connections of the network contain the RBF centers μ_j, obtained through unsupervised assimilation, followed by an output layer of weighting parameters, formed through supervised instruction. Training in the unsupervised mode is done without a pre-defined learning goal; input categorization and learning must be done using correlations within the input training data in contrast to feedback from a teacher or critic. For the RBFNN, the learning scheme essentially clusters the training inputs vectors and specifies where to position the RBF centers so that the desired response coverage is obtained. Thus, via unsupervised learning, the RBF center positions (the forward connections of Figure 10.2, μ_j) are chosen *a priori* and remain fixed throughout the establishment of the weighting coefficients (w_i).

The rearward connections, comprising the output layer of the network in Figure 10.2, specify the weighting (or regression) coefficients which are trained in a supervised fashion. Supervised means that the learning is based on comparison of the network output with the known "correct" answers.

For an RBFNN with a single layer of kernel functions, given that the basis function centers (μ_j) are fixed, the optimal weight array for the output connections which gives the best functional mapping can be found using the least squares normal equation developed in multiple linear regression theory. [5]

With all the parameters set, the fundamental mapping can then be written as:

$$f(\vec{x}) \approx \sum_{i=1}^{m} w_i g_i(\vec{x})$$ (10.4)

Thus, for an input vector $\vec{x}$, the solution $f(\vec{x})$ is a weighted linear summation of each RBF's response to $\vec{x}$. Kernel functions that have centers within the region of $\vec{x}$ will give the largest responses, whereas those farthest away will give negligible contributions to the series formed by Equation (10.4). Moreover, the kernel function responses ($g_i(x)$) will be bounded between 0 and 1 with the assigned weights (w_i) specifying the neuron's heights.

The following sections detail the procedures of RBF center selection and width estimation.

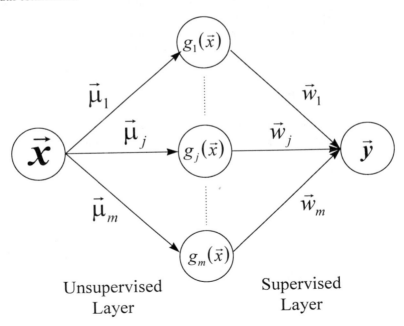

Figure 10.2 - RBFNN architecture.

Clustering and Center Selection

RBFNNs fix the center positions (μ_j's) and use simple multi-linear regression to obtain the optimal output weight array. To give this linear RBFNN more flexibility, a subset of RBF centers should be chosen which can best explain the variance in the dependent (or output) variables; in other words, centers are chosen so that the input space is adequately covered. Considering this, it seems natural to simply place a center over every input training point, $\mu_j = \mathbf{x}_j$ for j=1 to j=p (p designates the number of training cases). This choice would result in a network that "memorizes" all the training pairs and on occasion an added impropriety of exhibiting extremely poor generalization ability. The large number of neurons provides too many free parameters for the linear regression phase, thereby making the network oversensitive to the training set details. Furthermore, if the quantity of available training sets is large, then additional computational speed and memory problems will result, further making the RBF infeasible.

To counter these effects, the smallest subset of kernel function centers is selected that can perform the necessary mapping with a sufficient amount of accuracy. There have been many heuristics developed that can accomplish this subset selection process. Some start by adding one neuron at a time until a prespecified error goal is met. Others start with centers at all points in the initial input data and then make predictions on what subset will prove to be most efficacious.

RBF Width Estimation

The width parameter (σ in Figure 10.1) doesn't just control the shape of the RBF, it also specifies where and how much it overlaps with other RBFs spaced throughout the hypersurface. Physically, it determines the *shape of the response surface* (or the neuron's receptive field). Thus, if an extremely nonlinear problem is encountered, a greater amount of flexibility built into the surface construction will lead to a better performing network.

Recall the RBFs detailed in Equations 10.1-10.3. The parameter z introduces this "flexibility" by determining the type radial spread. Assuming that the network input vectors are n-dimensional and there are m neurons in the network, then z can take on the following forms (taken from [6]).

A. Each neuron has the same spherical size if r is constant,

$$g\left(\frac{\left\|\vec{x}_i - \bar{\mu}_j\right\|}{r}\right) \qquad (10.5)$$

B. Also, each neuron can have a different spherical size if $r = [r_1 \ r_2 \ \dots \ r_m]^T$,

$$g\left(\frac{\left\|\vec{x}_i - \vec{\mu}_j\right\|}{r_j}\right) \tag{10.6}$$

C. r can also be a vector in n-dimensional space giving each center the same ellipsoidal size with axes coincident with the basis vectors,

$$g\left(\left\|\left(diag(\vec{r})\right)^{-1}\left(\vec{x}_i - \vec{\mu}_j\right)\right\|\right) \tag{10.7}$$

D. If r is an n x n matrix **R**, then each center will have the same ellipsoidal size but with arbitrary axes,

$$g\left(\left\|R^{-1}\left(\vec{x}_i - \vec{\mu}_j\right)\right\|\right) \tag{10.8}$$

E. Finally, r can be a matrix of size n x m as $r = [r_1 \ r_2 \ \dots \ r_m]$ giving each center a different ellipsoidal size, with axes coincident with the basis vectors,

$$g\left(\left\|\left(diag(\vec{r}_j)\right)^{-1}\left(\vec{x}_i - \vec{\mu}_j\right)\right\|\right) \tag{10.9}$$

The choice of r for each neuron is usually left to random decision. There are some methods which allow for local "tuning" of the parameter, like those found in [4], but none account for the simultaneous evolution of the hypersurface. The next section provides an overview of what has been done in the past for these parameter estimations, with the following section outlining the genetic algorithm approach specific to the problem stated at the outset of the chapter.

PREVIOUS RBF PARAMETER SELECTION APPROACHES

Non-genetic Approaches

Non-genetic approaches to the problem of center (μ_j) selection typically separate a subset of centers from the training input patterns (x). A very crude method is to evolve one hidden unit at a time during the training processes, with the center being randomly selected from the input training vectors. The addition of hidden units would continue until some training error goal has been met. A more elegant approach uses a k-means clustering algorithm that groups together input patterns belonging to the same class region [4,7]. The vector mean positions of these regions are then used as the RBF center subset.

References [5,6] make use of statistical error prediction methods. For example, Bayesian Information Criterion and Generalized Cross Validation can be used to select input patterns from the training set that have the most probability of reducing the mapping error during the supervised training phase. This is done by looping through the training set and assigning scores to each input pattern, with the best score being selected. The selection process continues until the error prediction heuristic reaches a minimum. In other words, the addition of further RBF centers will not assist in the mapping error reduction.

Cascade-correlation can also be used with an RBF architecture for the evolution of one hidden unit at a time [8]. The process is actually a gradient descent procedure that places the RBF center in a position that reduces the current training error the most.

Non-genetic RBF width estimation procedures are limited. Most rely on a random search, sort of "brute force," method for width establishment. Reference [4] does provide local tuning procedures, but it doesn't consider the evolution of a whole network.

Genetic Approaches

The first major genetic invasion into neural network optimization can be found in the traditional backpropagation realm. The traditional backprop algorithm can incur long training times with subsequent poor generalization ability. To counter these effects, GA methods were developed that encoded individual layers of weights and optimized the network by de-linking (or pruning) those weights which were considered redundant. One such method was used successfully in [1].

Application of the GA to the RBFNN seemed to follow two paths: 1) Evolution based on search among whole networks and 2) Evolution of one neuron at a time. The methods of (1) would have a whole network be a single individual in the population. The GA would then choose among the various network structures (different centers, weights, and widths) as to which was the best. Those of method (2) would have a set of competing RBF units for one hidden unit addition (much like cascade correlation). Successive runs of the GA would then evolve the RBFNN one hidden unit at a time.

None of the aforementioned procedures give the desired flexibility necessary for the creation of higher order surfaces. To do this requires the simultaneous optimization of all RBF units within a single neural network. Whitehead and Choate contributed a seminal paper which attempted to do exactly this [2]. A cooperative-competitive algorithm was established in reference [2] that uses a population of strings to represent an RBFNN; not a population of RBFNNs. In other words, a single bit string was used to encode the center and width of one RBF. The algorithm attempted to evolve RBFs that would compete when doing the same job (in terms of the functional mapping), and implicitly share fitness (niching) when covering different parts of the overall functional region. In other words, if two RBFs are trying to cover the same area within the functional space, then they should compete, but if they're supplying individual contributions to the overall job of approximating the function, then they should form niches. The algorithm thus established mandated a fixed number of hidden layer neurons and a pre-chosen kernel type which applied to all neurons. Subsequent testing revealed that the GA approach offered 50-70% better performance than the best k-means approach.

The work of Whitehead and Choate [2] is significant in that it optimizes a single RBF network by evolving all the RBF units at once. As suggested earlier, anything that introduces more flexibility into the RBF construction can potentially improve performance (this was shown in [2]). The attempt of the current paper is to extend the viability of that statement by augmenting the work of [2]. Specifically, the individual kernel type choice, widths, and centers will all be allowed to vary during the evolution. The individual RBF neurons will have their parameters encoded as a binary bit string and the collection of these will comprise a single population.

GENETIC ENCODING

The space spanned by the center vector μ_i is defined on R^n, where n is the dimension of the input vector x_i. In [2], the inputs were scaled to occupy the unit hypercube $[0,1]^n$. This scaling forced the kernel centers (μ_i) to fall within the same unit hypercube. Encoding this arrangement required a translation of the fractional component along each dimension of the hypercube. The encoding scheme presented in the current effort does not place this restriction on center scaling. Centers are allowed to have components that have floating point numbers outside the range of $[0,1]$.

The overall objective is to evolve a set of basis function parameters for a layer of neurons. The parameters are the neuron's center, width, and basis function type. Assuming a binary alphabet $\{0,1\}$ with l bits of precision for the integer part and k bits for the floating point part of a center component, then $(\log(\text{max\_comp\_in\_pop} - 2) + l)^*n$ bits will completely encode a single center. The width can be encoded with $(l+k)$ bits of precision as well. Equations (10.1)-(10.3) give the various functions that a neuron can employ, thus, to encode the function type, will only require 2 bits (Gaussian is used for the redundant bit map). In summary, the parameters specifying a neuron are encoded in the following string form:

φ_i = (center)(width)(function type)

The bit length can be written as:

φ_i = {[log(max_comp_in_pop - 2)+*lJ**n}[*l*+*k*][2]

Therefore, a single generation will consist of a string population that completely encodes a single RBFNN architecture. *Successive generations will evolve the network as opposed to evolving a population of networks.*

FITNESS MEASURE

Reference [2] had reasoned that the relative significance of a neuron in the overall solution of Equation (4) could be associated with the size of weight assigned to it during the RBF training sequence. Subsequent work of [2] then found a fitness measure that promotes *competition* between neurons that are doing the same job and *cooperation* between those neurons supplying independent contributions to the approximation of f(x). As shown in [2], the competition component performs implicit niching; that is, niche sharing is actually accomplished without ever calculating a niche sharing function. The measure for doing this is stated below

$$\frac{\left|w_i\right|^{\beta}}{E\left(\left|w_{i'}\right|^{\beta}\right)}$$

where, w_i is the weight given to the neuron in a single generation, $E(\left|w_{i'}\right|)$ is the mean weight given to all the other neurons in the population, and β regulates the tradeoff between competition and cooperation ($1 < \beta < 2$).

In many function approximating simulations by the author, it was found that the assumption of a neuron's assigned weight determining its importance cannot always be justified. For example, considering the series approximation of Equation 10.4, a single neuron with an assigned weight could be replaced by two neurons at the same position with each possessing half the weight of the original neuron. Indeed, function approximation simulations have shown that the aforementioned fitness measure of [2] could lead to overcrowding of the RBF centers and a subsequent poor network formulation. For these reasons, the fitness measure specific to this chapter was developed in such a manner that allowed both an explicit and implicit calculation of fitness sharing (or niching) with a concomitant competitive factor, all of which are independent of a neuron's assigned weight. The following sub-sections detail the different aspects of the fitness measure.

A Neuron's Utility

With regard to the training data, a neuron's importance can be determined by the amount of data points for which it is actually contributing significant responses . If there are p training patterns, then the number of times an RBF neuron responds with a value greater than a preset utility factor (α) would be a direct indication of how much the neuron is doing to approximate $f(x)$. Thus, a utility measure can be stated as:

For,

$$\left\{ g_i\left(\vec{x}_j\right) \right\}_{j=1,p}$$

$$UM = (\text{number of } g_i\text{'s} > \alpha)$$

It was found in subsequent studies that a value of $\alpha=0.5$ worked best for most trial problems.

Overlap Measure

Simply awarding neurons for covering data will certainly lead to a crowding of neurons around the most dense portion of the data hull. Therefore, there needs to be a measure which will award neuron separation. This can be defined by understanding the nature of the RBF parameters. Using the center positions and widths of two neurons, a measure can be written in terms of an overlap parameter:

$$\lambda = E\left(\frac{\left\| \vec{\mu}_i - \vec{\mu}_j \right\|}{r_i + r_j} \right)$$

Here, $E(\ )$ defines the mean separation of neuron i from all other neurons for which j does not equal i. λ values greater than unity specify separation of internal regions with respect to the neurons widths, values equal to unity imply touching edges of the regions, and values less than 1 signify overlap of internal regions. It is obvious that such a measure would award those centers which sought to increase their respective λ's.

The two foregoing measures can be written in collective form as follows:

$$\text{Fitness Measure} = (W_1 * UM + W_2 * \lambda)$$

The choice of the W_i weighting factors can be determined by an empirical study for the specific problem under consideration. The final section of this chapter subjects this measure to a highly nonlinear approximation problem.

APPLICATION TO SSME PLUME SPECTRAL ANALYSIS

For years, researchers at NASA's Marshall Space Flight Center (MSFC) have been filming developmental tests of the SSME. Subsequent analysis of film, which involved major engine failure incidents, revealed a consistent feature common to many of the breakdowns. In eight of twenty-seven failure events observed, a visible discharge of some substance was seen in the plume prior to engine failure [9,10]. These discharges ranged from extreme plume flashes to small regional streaks. This discovery lead to the thought that if it is possible to visually detect anomalous events with the naked eye, then perhaps it would be possible to discover anomalous events below the "visible threshold" (i.e., infrared, ultraviolet, etc.) that would be indicative of an imminent failure. This would be tantamount to real-time engine health monitoring via the quantitative analysis of the SSME exhaust plume spectral data.

The idea of extracting information about a system from its electromagnetic (EM) emission is certainly not new. Consider a piece of nickel being held over an open flame; at some point in the heating process if the flame is hot enough, the molecular structure of the nickel will become so excited that it begins to radiate. As shown in Figure 10.3, this radiation, being electromagnetic in nature, can be represented as a spectrum of EM intensity versus transition wavelength. Therefore, with a spectrometric detector some distance away during the nickel experiment, it would be a simple task to identify the metal which is being burned.

Individual spectra for other elements vary in complexity. Some have a few atomic transitions (peaks) while others have many. Three germane points of importance to the current effort result from Figure 10.3 and the associated radiation physics: 1) *every element has its own "spectral signature,"* 2) *the emission will contain atomic transitions at wavelengths which may not be part of the visible spectrum, and* 3) *the **intensity** of the emission is a function of the **quantity** of emitting matter present in addition to the system temperature and other quantum variables.*

Rocket plumes are emissive events subject to the same physics (with more complications, of course) as burning nickel over an open flame. The Optical Plume Anomaly Detection (or OPAD) program was initiated by researchers at MSFC as an effort to take advantage of the wealth of information contained in the exhaust plume of a rocket engine. The initial idea was to identify anomalous spectral events which were consistent with known mechanical failures and then use them as templates in the health monitoring of future engine tests (ground or in-flight). This could then be coupled with the anomalous events found in the vibrational and other sensor data to determine the overall state, or health, of the engine.

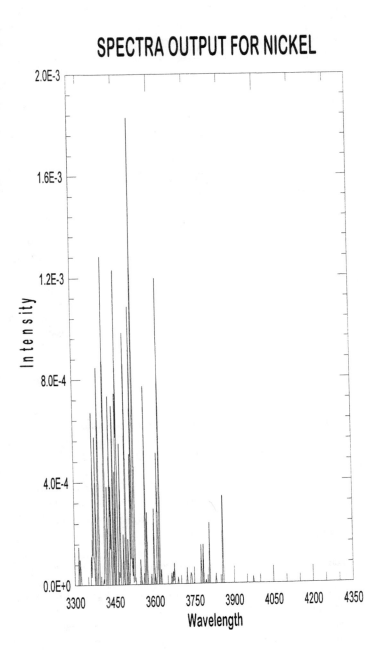

Figure 10.3 - Example emission spectra for nickel.

Using plume spectral acquisition instrumentation developed specifically for OPAD and mounted on the SSME Technology Test Bed (TTB) at MSFC, the process of building a cumulative database of the spectral templates began [9]. Researchers attempted to catalog the various spectral forms associated with changes in the SSME's operation (i.e., changes in oxidizer/fuel ratio, engine startup transients, etc.) so that a baseline of "expected spectral signatures" would be established. The "template idea," however, soon gave way to even more ambitious goals as a result of some initial findings in the TTB experimental program. The spectral data from one test in particular revealed a major occurrence of a metallic species which was indigenous to the SSME preburner faceplate. An even closer evaluation of the amount of metallic species present versus time showed an initial erosive event of the metal followed by numerous other anomalous emissions, all leading up to an engine-threatening erosion of the faceplate. This meant that anomalous events could be predicted.

As a result of these findings, the focus of the researchers turned to not only anomaly detection but also metal quantification. In other words, health monitoring now involved the simultaneous tasks of anomaly detection and determination of the severity of the anomaly. This meant that the free atom densities of all the metals of interest within the engine would have to be predicted for every temporal scan taken by the instruments. The metal quantification process would essentially give metal concentration versus time. Spikes in this time trace would then be indicative of a metal erosion. Figure 10.4 summarizes the three major objectives of the health monitoring system.

Any spectrum obtained with the OPAD instrumentation is composed of three components: 1) a dominant OH component which arises from the burning of dissociated hydrogen radicals, 2) a background noise component caused by the scattering of background light, and 3) a metallic component, if indeed there is one, which would be indicative of a metal erosion. Thus, the quantification of a metal erosion and the subsequent identification of any anomalies requires a spectral "cleaning" procedure followed by an evaluation of the plume metallic state. In other words, methods for removing the OH and background components of the spectrum would need to be employed so that the underlying metallic component could be seen. Then, the metal quantities would have to be ascertained from the remaining metallic component.

For a given spectrum, ascertaining the metallic quantity could only be done through two methods: 1) by comparing the spectrum to past spectra obtained from plume seeding tests, or 2) using a theoretical model that emulates the emissive nature of the plume. The first option is plagued by an inability to precisely measure the erosion and survivability rates of the inserted species. Thus, there would exist plume spectra to compare to, but the metallic content associated with the spectra would be in error. Moreover, it would not be cost effective to run the SSME through all the possible metallic seeding combinations. For these reasons, option two was selected.

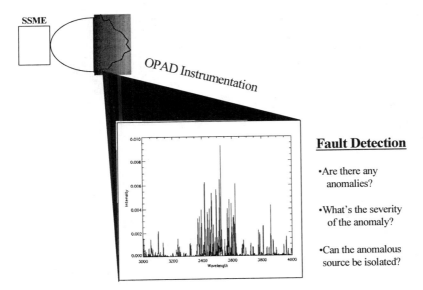

Figure 10.4 - Overview of tripartite monitoring scheme.

Researchers at Vanderbilt University and AEDC developed a theoretical plume model known as SPECTRA [11,12]. The forward operation of the SPECTRA code involves the calculation of a theoretical plume spectrum from a pre-defined set of metal concentrations and flow parameters. The reverse operation of SPECTRA (obtaining the metallic components which made up the spectrum) cannot be written in a numerically convenient analytical form because of the almost insurmountable mathematics involved. This, therefore, mandates that the SPECTRA code be applied in an iterative manner until it converges with the spectrum obtained from the OPAD instrumentation. The set of SPECTRA input parameters which produced this convergence would then specify the current metallic state of the plume. Operating this iterative sequence is computationally exhaustive and precludes its use in real-time health monitoring systems. For this reason, neural network techniques have been investigated that accomplish this fundamental task in an expeditious manner. Described herein is a radial basis function neural network (RBFNN) architecture that models the "inverse" operation of the SPECTRA code and allows for real-time SSME anomaly identification and quantification via plume spectral assessment.

GA/RBFNN TESTING AND RESULTS

The parameters typically predicted from spectral scans are the plume temperature, the number densities for the metallic species present, and the broadening parameter. Number density and broadening parameter can be used to obtain the metallic species concentration. To test the viability of the GA hybrid approach to the plume emission problem, spectral data was generated for the element Chromium. The data represented example input (spectra) to output (temperature, number density, and broadening parameter) RBFNN variables. Three scenarios were investigated: 1) Training the RBFNN using the random RBF center selection method; 2) Training with the Bayesian Information Criterion (BIC) selection method; and 3) Using the GA RBF center, basis function, and width evolution method. Having trained three separate RBFNNs using the aforementioned methods, a separate data set was created for the purpose of testing. This data was not seen by the RBFNNs during training and should provide an excellent measure of the RBFNN's generalization ability. Table 10.1 below shows the resulting performance of the three methods. The mean percentage error on the test data is given as well as the standard deviation error for the RBFNN predictions.

Note, the same test data was used in all three cases. Preprocessing and scaling of this data could improve or degrade each of the networks performances. However, by using sound preprocessing schemes and maintaining the same data set it is hoped that this bias can be removed, thereby allowing comparisons between the methods to be stated. Table 10.1 reveals that the GA hybrid approach is much better than the random scheme and at least as good as the BIC scheme. The BIC scheme is an excellent selection method for nonlinear problems so, considering the results, it is apparent that the GA hybrid is just as effective at finding good solutions.

Table 10.1 - Resulting prediction errors for three methods.

	Mean Percentage Error			Standard Deviation Error		
	Temp	Num Den	BP	Temp	Num Den	BP
Random	15.38	88.47	375	0.3716	9.554	4.688
BIC	1.69	76.85	169.52	0.007	0.249	0.083
GA	5.49	34.94	130.04	0.013	0.365	0.080

A problem was noted with the particular fitness function detailed earlier. The performance of the RBFNN after GA center evolution was extremely sensitive to the weighting factors. Moreover, during the GA center, evolution the RBFNN performance error oscillated between suitable and unsuitable values. This was most likely caused by the method in which fitness sharing was implemented. To this end, more work will need to be done.

Finally, the goal was to evolve RBF basis functions that competed, yet cooperated, to approximate the mapping between the spectrum and the plume physical parameters. Thus, it was hoped that the average population fitness would increase and eventually match the fitness of the best performer. Figure 10.5 below shows the fitness performances versus the generation number for one particular GA run. Note, on the average, the fitness continues to increase for both plots, but the average population fitness doesn't approach the best fitness like it should. Once again, this is most likely caused by the niching function and its establishment. Notwithstanding this reservation, the GA thus established was able to match the performance of the BIC scheme. This certainly reveals promise in the method and further research would most likely prove fruitful.

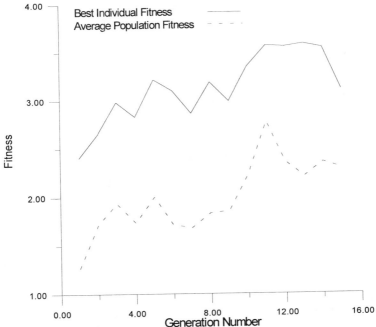

Figure 10.5 - Fitness performance versus generation number.

CONCLUSIONS

A hybrid RBFNN and GA was developed that allowed for the evolution of the internal RBF parameters in such a fashion that contributed to the overall reduction of RBFNN mapping error. Considering the physical nature of the RBFNN, a fitness function was established that implemented both competition and cooperation between population members. The hybrid GA/RBFNN was tested on actual SSME plume spectral data. The results showed that the hybrid scheme is much better than the random scheme and just as good as the current BIC scheme in use. Improvements can be made to the niching implementation that may attenuate the undesirable sensitivity and oscillations seen in the evolution. Finally, it should be noted that this functional mapping is highly nonlinear and the performance of the GA is very encouraging.

REFERENCES

[1] Benzing, D. A., Whitaker, K. W., & Krishnakumar, K. (1997). Experimental verification of neural network-based SSME anomaly detection. *Pro-*

ceedings of the 33rd AIAA/ASME/SAE/ASEE Joint Propulsion Conference, AIAA-97-2903, Seattle, WA,   July 6-9.

[2] Whitehead, B. A. & Choate, T. D. (1996). Cooperative-competitive genetic evolution of the radial basis function centers and widths for time series prediction. *IEEE Transactions On Neural Networks,* **7**(4).

[3] Mackey, M. C. & Glass, L. (1977). Oscillation and chaos in physiological control systems. *Science,* **197**, 287-289.

[4] Moody, J. & Darken, C. J. (1989). Fast learning in networks of locally-tuned processing units. *Neural Computation,* **1**, 281-294.

[5] Orr, M. J. L. (1996). *Regularisation in the selection of radial basis function centres,* Centre for Cognitive Science, University of Edinburgh, 2, Buccleuch Place, Edinburgh EH8 9LW, UK.

[6] Orr, M. J. L. (1996). *MATLAB routines for subset selection and ridge regression in linear neural networks.* Centre for Cognitive Science, Edinburgh University, Scotland, UK, April.

[7] Musavi, M. T., Ahmed, W., Chan, K. H., Faris, K. B., & Hummels, D. M. (1992). On the training of radial basis function classifiers. *Neural Networks,* **5**, 595-603.

[8] Fahlman, A. E. & Lebiere, C. (1991). The cascade-correlation learning architechure. *Advances in neural processing systems 2,* (Eds.)R.P. Lippmann, J.E. Moody, & D.S. Touretzky, San Francisco: Morgan Kauffmann Publishers, 524-532.

[9] Powers, W. T., Cooper, A. E., & Wallace, T. L. (1992). OPAD status report: Investigation of SSME component erosion. *SAE Paper 92-1030.*

[10] Cikanek, H. A. III (1987). Failure characteristics of space shuttle main engine failures. *Proceedings of the Joint Propulsion Conference,* June.

[11] Powers, W. T., Cooper, A. E., & Wallace, T. L. (1995). Validation of UV-VIS atomic spectral model for quantitative prediction of number density, temperature, and broadening parameter. *1995 JANNAF Propulsion Systems Hazards Subcommittee,* Marshall Space Flight Center, AL, October.

[12] Wallace, T. L., Powers, W. T., & Cooper, A.E. (1993). Simulation of UV atomic radiation for application in exhaust plume spectrometry. *Proceedings of the 29th Joint Propulsion Conference and Exhibit,* AIAA Paper 93-2512, Monterey, CA.

CHAPTER 11

TUNING BAMA OPTIMIZED RECURRENT NEURAL NETWORKS USING GENETIC ALGORITHMS

K. Nishita
Graduate Student
Department of Aerospace Engineering and Mechanics
University of Alabama
Box 870280
Tuscaloosa, AL 35487-0280
e-mail: knishita@eng.ua.edu

ABSTRACT

Genetic Algorithms (GAs) have demonstrated highly efficient and robust in optimization capabilities in various fields. This chapter discusses the use of a simple GA for enhancing the performance of a Bama Optimized Recurrent Neural Network (BORN). In order to measure the performance of BORN with a GA, two different studies are considered. First, the results produced by the GA optimized BORN were compared to results produced using the BORN algorithm alone. Second, a GA was used to alter the connectivities in a BORN adaptive unit to enhance the performance of the BORN algorithm. The advantage of this method is that a GA can improve the performance of the BORN by changing only the NN connections. In other words, since previously trained BORN weights are not changed, the BORN is capable of retaining its memory. Results show that a GA enhanced BORN's adaptive capabilities with a high degree of accuracy.

INTRODUCTION

Background of GA Application in NN

Genetic Algorithms (GA) have been demonstrated highly efficient and robust optimization capabilities in various fields. In recent years, many researchers have sought to apply GAs in the area of neural networks (NNs). NNs have been successfully adapted to the number of practical engineering problems such as controllers, classifiers, etc. NNs have great potential in representing arbitrary surfaces and can be easily implemented. Various schemes for combining GA and NN have been proposed and tested to optimize NN performance in recent years. Montana and Davis [1] reported the first successful results of a relatively large NN (500 connections) optimization by GA. However, the GA used by Montana and Davis was different in several ways from the more traditional GA described by Goldberg [2]. Belew [3] used a GA to select initial NN weights and trained NN by using a gradient descent method. His results showed GA and NN combined method performed better than instances in which only a GA or a NN gradient decent method were used. Anderson [4] applied a GA for NN reinforcement learning problems in which desired outputs are not available for training purposes. He used his system to successfully control an inverted pendulum. Anderson also compared these results with his adaptive critic controller and found that GA for NN reinforcement learning was competitive with NN training method.

Recently, the most considerable attention has been focused on recurrent neural networks. The effectiveness of recurrent neural networks for the identification and control of nonlinear dynamic systems has been demonstrated by Werbos [6] and Narendra [7] et al. Wieland [5] also used GA to optimize recurrent NN for two poles jointed inverted pendulum. He used a simple GA [2] to optimize a recurrent NN controller to successfully control this difficult, highly nonlinear problem. For nonlinear system controllers such as an inverted pendulum, the effectiveness of the control system must be gauged not only by the speed of which the goal is achieved, but also by the stability and the robustness. KrishnaKumar [8,9,10] et al. demonstrated recurrent neural network stability and robustness by using the method of network connection optimization (Bama Optimized Recurrent Neural Network). Bama Optimized Recurrent Neural Network (BORN) is a back-propagation neural network (BPNN) which has the capabilities of network connectivities optimization. This optimization algorithm was first introduced by KrishnaKumar [8] for static neural networks and adapted for recurrent neural networks in [9]. BORN has demonstrated high robustness and generalization capability in various system control areas [10]. However, a disadvantage of this algorithm is that once the BORN algorithm is performed, connectivities cannot be altered without re-initializing its weights and connections. This chapter explores control capabilities of the BORN algorithm and considers a mechanism to address this disadvantage by using a GA.

Study Objective

The objective of the work presented in this chapter is to employ a GA in a BORN algorithm to alter the NN connections after the BORN algorithm execution. In order to fulfill this objective, the following studies will be undertaken:

- Perform BPNN weight selections and connectivities optimizations using a GA and compare the results with classic BORN algorithm results.

- Once the BORN algorithm is executed, the network connectivities cannot be adapted in BORN. In order to enhance a BORN, a GA will be used as a BORN connectivities adaptive unit for further optimal weight and connection selections. The GA adaptive unit will be executed only when BORN detects the failure to achieve its goals due to sudden changes in input patterns.

Chapter Outline

The reminder of this chapter is organized into four sections. First, BORN algorithms are introduced and the details of the mathematical model for a BORN are emphasized. Second, the details of BP Recurrent Neural Network weights selections and connectivities optimizations by GA are presented. The results of the GA optimization performance will be presented at the end of this section. Third, the concepts of GA-BORN connectivities adaptive unit are presented. Finally, the GA-BORN results are compared to those of a non-adaptive BORN.

SIMPLE GA AND BORN ALGORITHM

This section introduces simple GA and BORN algorithm. For the sake of completeness, the conventional backpropagation recurrent neural network algorithm is also presented here.

Simple GA

GAs are optimization algorithms inspired by biological evolution [2]. They have been shown to be very effective at function optimization, efficiently search large and complex spaces to find near global optima. A simple GA uses three operators in its quest for improved solutions: reproduction, crossover, and mutation. These operators are implemented through the performance of the basic tasks of copying binary strings, exchanging portions of strings, and generating random numbers, respectively.

Reproduction is a process in which strings with high performance indexes receive accordingly large number of copies in the new population. For instance, in roulette wheel reproduction, the strings are given a number of copies that are proportional to their fitness. The probability of reproduction selection is defined as in Equation (11.1).

$$P_{select} = \frac{f_i}{\sum f} \qquad (11.1)$$

where

P_{select} = Probability of a string being reproduced, and
f_i = fitness values of an individual string.

Reproduction drives a population toward highly fit regions of the search space.

Crossover provides a mechanism of information exchanges between high performance strings. Crossover can be achieved in three steps:

1. Select two new strings from the mating pool of strings that were produced by reproduction.
2. Select a position for the crossing site.
3. Exchange all characters following the crossing site.

An example of a crossover is shown Figure 11.1. The binary coded strings A and B of length 10 are crossed at the third position. New strings A' and B' are produced.

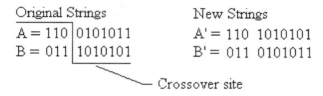

Figure 11.1 - Example of GA crossover.

Mutation enhances a GA's ability to find near optimal solutions by providing a mechanism to insert missing genetic material into the population. Mutation consists of the occasional alteration of a value at the particular string position. This procedure insures against the loss of a particular value at any bit position.

Together, reproduction, crossover, and mutation provide the ingredients necessary for an effective GA. This simple GA model is employed in the current study.

Recurrent NN

Recurrent Neural Networks (RNNs) are experiencing an increasing popularity because of their inherent dynamic nature. In RNNs, individual neurons are fed back as inputs to other neurons. The general structure of an RNN with BP learning is shown in Figure 11.2. In this figure, the circles represent the neurons of the RNN and the arrows are RNN connections. Each connection has its own strength, commonly called a weight. In typical BP learning, weights are adjusted in order to obtain the desired output values from the RNN input. RNN input can be any crisp values, which are related to the RNN output values. RNN input values are usually obtained from the environmental state, and the outputs are the predicted action or consequence of the input. Therefore, like most NNs, RNNs attempt to capture the relationship between input and output values, inherent in a particular problem. As shown in Figure 11.2, an RNN neuron has connections from every other neuron to its left at time period t. Also, every neuron has connections from itself and from every other neuron to its right at time t+dt. Here, time has no meaning in the physical environment; it is used exclusively to mark iterations through an NN learning cycle.

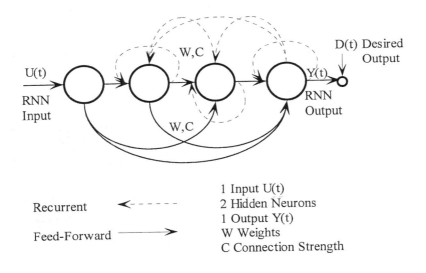

Figure 11.2 - Recurrent NN with BP learning structure.

The operation of RNNs with BP learning consists of two parts: (1) a forward pass and (2) a backward pass. The primary role of the forward pass is to predict an output response from a given input. The outputs are a definite function of the input. When an individual neuron receives an input, the input goes through an activation function within the neuron and generates a neuron output. The activation function can take many forms. Generally, it is a nonlinear function, but its only true limitation is that it must be a differentiable function. In this study, sigmoidal functions (Figure 11.3), are used. Equations 11.2 to 11.5 are necessary to accomplish a forward pass. Equation 11.2 represents the inputs U, which are received by the RNN. These inputs are multiplied by weights in 11.3 and outputs are processed by Φ_i in Equations 11.4 and 11.5. The $\Phi_i(t)$ is the activation function. The effect of a bias in the neurons is achieved by assuming that the first input is always unity and is connected to all the other neurons.

$$x_i\ (t) = U_i\ (t) \qquad\qquad 1 \le i \le m, \qquad\qquad (11.2)$$

$$Net_i(t) = \sum_{j=1}^{i-1} W_{ij}\ x_j(t) + \sum_{j=i}^{N} W_{ij}\ x_j(t-1) \qquad m+1 \le i \le N,$$

$$(11.3)$$

$$x_i(t) = \Phi\big(Net_i(t)\big) \qquad m+1 \le i \le N, \qquad\qquad (11.4)$$

$$Y_i(t) = x_{i+m+h}(t) \qquad 1 \le i \le n, \qquad\qquad (11.5)$$

where

t	= current time frame,
$t-1$	= previous time frame,
$U(t)$	= net inputs,
$x(t)$	= neuronal activations,
$Y(t)$	= net output,
$\Phi_i(t)$	= activation function,
W_{ij}	= weight connecting the ith neuron to the jth neuron,
m	= number of inputs,
h	= number of hidden neurons,
n	= number of outputs,
N	= total number of neurons $(m+h+n)$.

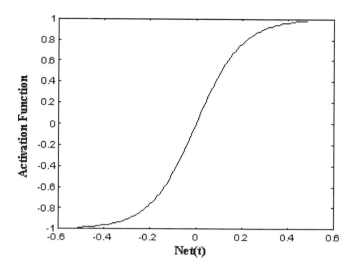

Figure 11.3 - Sigmoidal activation function $\Phi_i \left(Net(t) \right) = \dfrac{1 - e^{-10 \cdot Net(t)}}{1 + e^{-10 \cdot Net(t)}}$.

The second fundamental operation, the backward pass, is where the learning or adaptation occurs. In a backward pass, errors associated with the RNN's performance are used to adjust the weights associated with the connections. Here, the errors are often a sum of squared error between the desired output (for the given input) and the output actually produced by the RNN. This adaptation (BP learning in this chapter) implies a modification of RNN structure and its parameters based on repeated exposure (epoch) to the environment, or from input-output pairs collected from the environment. Equations 11.2 to 11.6 are necessary to accomplish a backward pass. RNN learning involves an evaluation of RNN performance. The performance is measured via an error that is computed using Equation 11.6. The sum of the differences of RNN outputs ($Y_f(t)$) and RNN desired outputs ($d_f(t)$) is squared:

$$E_i(t) = \frac{1}{2} \sum_{i=1}^{n} \left(Y_i(t) - d_i(t) \right)^2 \qquad (11.6)$$

Based on this performance measure, RNN weights are altered in the direction that is most likely to reduce the RNN error $E_f(t)$. This alteration is driven by a derivative-based algorithm. The error gradient is derived from Equation (11.6). The error gradient are computed according to

$$\frac{\partial E}{\partial Y_i(t)} = Y_i(t) - d_i(t), \qquad 1 \leq i \leq n \quad (11.7)$$

where $d(t)$ = desired (target) output.

Ordered derivatives are a special case of the chain rule only applicable for an ordered system where the values can be computed one by one in the order of E_1, E_2, ... E_i. By using the definitions of ordered derivatives presented by Werbos [6], the error derivatives for each neuron are computed according to

$$\frac{\partial^+ E(t+1)}{\partial X_i(t)} = \frac{\partial E}{\partial X_i(t)} + \sum_{j=i+1}^{N} \frac{\partial^+ E}{\partial X_j(t)} \frac{\partial X_j(t)}{\partial X_i(t)} + \sum_{j=m+1}^{i} \frac{\partial^+ E}{\partial X_j(t+1)} \frac{\partial X_j(t+1)}{\partial X_i(t)},$$

$$N \geq i \geq 1 \tag{11.8}$$

where

$$\frac{\partial E}{\partial X_i(t)} = \begin{bmatrix} 0, & 1 \leq i \leq m+h \\ \frac{\partial E}{\partial Y_{i-(m+h)}(t)}, & m+h \leq i \leq N \end{bmatrix}, \qquad \frac{\partial X_j(t)}{\partial X_i(t)} = \Phi'_j\big(Net_j(t)\big)\, W_{ji}$$

$$\tag{11.9,10}$$

The ordered derivatives for the weights can be calculated as

$$\frac{\partial^+ E}{\partial W_{ij}} = \frac{\partial^+ E}{\partial X_i(t)}\, \Phi'\big(Net_i(t)\big)\, X_j,$$

$$1 \leq i \leq N \text{ and } 1 \leq j \leq N \tag{11.11}$$

And, the weights are updated according to

$$W_{ij} = W_{ij} - \varepsilon \frac{\partial^+ E}{\partial W_{ij}}, \quad 1 \leq i \leq N \text{ and } 1 \leq j \leq N \tag{11.12}$$

where ε is a learning rate, which is used for RNN learning speed adjustments. The larger the value of the learning rate, the faster adaptation occurs. However, if the learning rate is too large, the RNN can diverge, leading to increasing errors and erroneous results.   On the other hand, the smaller the value of the learning rate, the slower adaptation tends to be, but the more likely the RNN is to converge.

In RNNs, Equations 11.2 to 11.12 are applied repeatedly until the RNN can produce the desired outputs within a given tolerance. This tolerance is set by the RNN user.

As mentioned earlier in this section, different NNs have very different structures and distinguishing characteristics. Therefore, NN users must select a NN that is most suitable to the particular problem they wish to solve.

BORN is an RNN connection optimization algorithm, which can eliminate unnecessary RNN connections. A brief explanation is provided below. In order to use the BORN algorithm in RNNs with BP learning, a new function $g(C_{ij})$ must be introduced into the RNN connections. This function operates on the connections between neurons. If $g(C_{ij})=1.0$, a connection between the ith and j^{th} neurons is present. If, however, $g(C_{ij})=0.0$, the connection is not present. In order to apply the BORN algorithm in RNNs, Equation 11.3 must be changed to

$$Net_i(t) = \sum_{j=1}^{i-1} W_{ij}\, g(C_{ij})\, x_j(t) + \sum_{j=i}^{N} W_{ij}\, g(C_{ij})\, x_j(t-1),$$
$$m+1 \le i \le N \qquad\qquad (11.13)$$

where $\quad g(C_{ij}) = \dfrac{1}{1+e^{-\alpha \cdot C_{ij}}},$

where α is a constant, typically set to a value greater than 10.0. For sufficiently large α, $g(C_{ij}) \cong 1.0$ for $C_{ij} > 0$, and $g(C_{ij}) \cong 0.0$ for $C_{ij} < 0$.

If the connection strength, C_{ij} , is set to 0.0 at the beginning of the training, and if BP is applied to update C_{ij}, $g(C_{ij})$ will approach either 1.0 or 0.0. Now, the error derivative can be expressed as

$$\frac{\partial X_j(t)}{\partial X_i(t)} = \Phi'_j\big(Net_j(t)\big)\, W_{ji}\, g(C_{ji}). \qquad\qquad (11.14)$$

For the RNN, the error gradients for the weights are expressed as

$$\frac{\partial^+ E}{\partial W_{ij}} = \frac{\partial^+ E}{\partial X_i(t)}\, \Phi'_i\big(Net_i(t)\big)\, g(C_{ij}) X_j(t) \qquad\qquad (11.15)$$

and $\dfrac{\partial^+ E}{\partial C_{ij}} = \dfrac{\partial^+ E}{\partial X_i(t)} \, \Phi'_i\big(Net_i(t)\big) \; W_{ij} \; X_j(t) \; g'\big(C_{ij}\big)$ (11.16)

The weight and connection updates are expressed as

$$ W_{ij} = W_{ij} - \varepsilon_w \frac{\partial^+ E}{\partial W_{ij}} \quad \text{and} \quad C_{ij} = C_{ij} - \varepsilon_c \frac{\partial^+ E}{\partial C_{ij}}. \qquad (11.17,18) $$

For RNN training, Equations 11.13 to 11.18 are applied repeatedly until the RNN can produce the desired outputs within some pre-defined tolerance on the RNN error. Further details of the BORN algorithm and example applications can be found in Reference [12].

NN CONNECTIVITIES OPTIMIZATION WITH GA

This section performs BP Neural Network weight selection and connectivity optimizations by GA and compares the results with BORN algorithm results. The purpose of this study is to compare the stability and robustness of the BORN algorithm and GA. Simple GA [2] was used to verify the performance of the Neural Network with GA optimization.

Decode and Fitness Function for GA-NN

In order to optimize NN connections by GA, a decode routine and a fitness function must be developed. The number of NN connections is defined as the sum of the number of NN inputs, hidden neurons, and outputs. Since, in this chapter, the NN connections are represented in binary form (such as connected is 1 and disconnected is 0), this can be translated to GA as a 1-bit multi-parameter problem. Therefore, the decode function is applied such as "if an allele is 1, then the NN connection exists; otherwise no connection" for the number of connections, which is essentially the same as the chromosome length. Listing 11.1 is the complete C language listing for decode function used in this study. The objective of the GA is to maximize the fitness function, f, defined in Equation 11.19.

$$f = \frac{1}{\Big(NN\ Output - Desired\ Trajectory\Big)^{2}} \qquad (11.19)$$

```
void decode(chromosome *chrom, int *lchrom)
{
int i, j, counter;
counter = 0;
for(i=(NN_Num_Input+ NN_Num_Hidden+ NN_Num_Output);i>=1;--i){
for(j=( NN_Num_Input+ NN_Num_Hidden+ NN_Num_Output);j>=1;--j){
if ( chrom[counter] ){ ConnectionMatrix[i][j] = 1; } //Connection Exists
else { ConnectionMatrix[i][j] = 0;} //No Connection
++counter;}}
return;
}
```

Listing 11.1 - Decode function for GA-NN connection optimization problem.

GA-NN Performance Evaluation

The dynamic system presented in [7] is used for the performance evaluation for both BORN and GA connectitivities optimization methods. The NN that will be optimized by BORN has three inputs, five hidden neurons, and one output. This NN is trained with learning rate of 0.01 for 75 epochs and then BORN algorithm is executed for 25 epochs. The NN optimized by GA has the same structure as the BORN optimized NN. In order to optimize NN connections by GA, GA parameters are set to 20 populations, 9 strings, 0.8 probability of crossover, and 0.01 probability of mutation for 100 generations.

Figure 11.4 shows the performance comparison for both GA-NN and BORN. Both of the methods predicted the response of the dynamic system well. Although, from Figure 11.5, it is clear that the GA-NN found a near optimal solution, BORN algorithm generates better performance for this training case in terms of NN output error. Tables 11.1 and 11.2 are the comparison of NN connections between BORN and GA-NN. These tables are labeled as o - no connection (there are no connections between input units), x - connected, and d – disconnected by optimization. GA-NN determined four more unimportant connections than BORN. Figure 11.6 shows the simulation results of optimized BORN and GA-NN using different data sets. The data sets were generated according to the following equation:

$$u(k)=sin(2\pi k/150 + 0.05) \qquad\qquad (11.20)$$

Figure 11.6 indicates that both NN showed highly robustness for frequency and phase changes from the trained data sets. Additionally, Figures 11.7 to 11.10 further demonstrate the effectiveness of both NNs in modeling the dynamic system considered here.

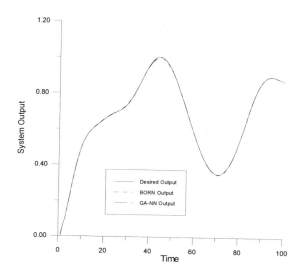

Figure 11.4 - System output vs. time plot.

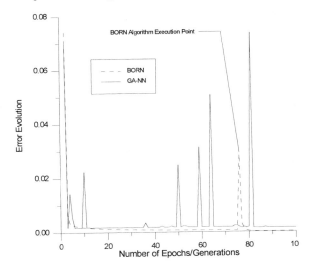

Figure 11.5 - Error evolution plot.

Table 11.1 - NN Weights by BORN. (28 disconnections)

o	O	o	o	o	o	o	o	o
o	O	o	o	o	o	o	o	o
o	O	o	o	o	o	o	o	o
d	D	d	d	d	d	d	d	d
d	X	d	d	d	x	d	x	x
d	X	x	x	d	x	d	x	d
x	X	d	x	x	x	d	x	d
x	D	d	x	d	x	x	x	x
d	X	x	d	d	x	x	d	x

Table 11.2 – NN Weights by GA. (32 disconnections)

o	o	O	o	o	o	o	o	o
o	o	O	o	o	o	o	o	o
o	o	O	o	o	o	o	o	o
d	x	D	x	d	d	x	x	x
d	d	X	x	d	d	d	d	d
d	x	X	x	d	x	x	d	d
x	x	D	d	d	x	d	d	x
d	d	D	d	d	x	d	d	d
d	x	X	d	d	x	d	x	x

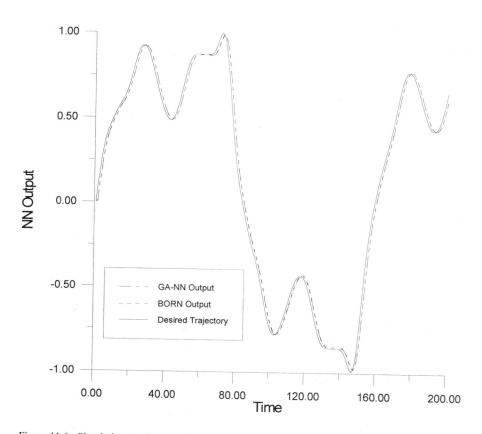

Figure 11.6 - Simulation result comparison.

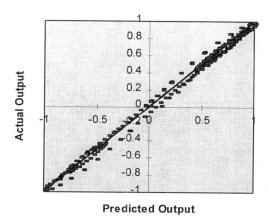

Figure 11.7 – 45° plot for BORN.

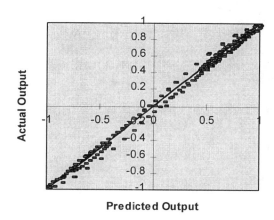

Figure 11.8 - 45° plot for GA-NN.

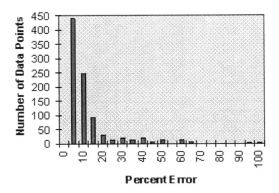

Figure 11.9 - Percent error distribution for BORN.

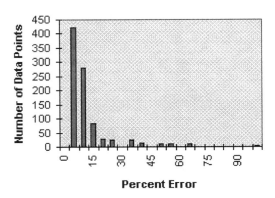

Figure 11.10 - Percent error distribution for GA-NN.

GA-BORN ADAPTIVE UNIT

This section introduces the concept of GA adaptive unit for BORN. Both a simple architecture for the GA-BORN adaptive system and simulation results of a hydrocyclone modeling problem are presented here.

GA-BORN Adaptive System Architecture

In the classic application of the BORN algorithm, once the BORN algorithm is executed, the NN connectivities cannot be changed. In order to alter the NN connectivities after BORN optimization, a GA is implemented as an adaptive unit for further optimal connection selections. Figure 11.11 shows the simple GA-BORN Adaptive System Architecture Diagram. In this figure, an NN is trained as a general controller with BORN optimization for the system. Once simulation starts, the Performance Estimator checks the NN controller output error. The GA adaptive unit should be executed only when the Performance Estimator detects the unacceptable NN output for the system due to sudden changes in input patterns, or the environment. For example, the NN is trained and optimized by BORN for input U=sin x as a general controller. In the case of input frequency and phase changed to $U = \sin(2x + \pi/2)$ for some reason during operation, the GA adaptive unit will look for more optimal connectivities for the NN controller. The advantage of this method is that since only the NN connectivities changed, and NN weights, which is the memory part of the NN, will not be changed by GA, the NN can retain its memory which was previously optimized by BORN. In other words, in order to regain the original NN after the GA optimization, the previous connectivities can be applied without further NN training. Decode and fitness functions are essentially the same as described in previously in this chapter.

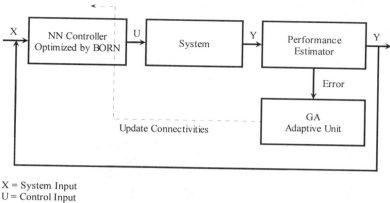

X = System Input
U = Control Input
Y = System Output
Error = $(Y - \text{Desired } Y)^2$

Figure 11.11 GA-BORN adaptive system architecture diagram.

Hydrocyclone Modeling Problem

Hydrocyclone modeling problem is used for the performance evaluation for GA-BORN adaptive unit. Hydrocyclones are commonly used for separating slurries in the mineral processing industry [11], and are used extensively to per-form separations in other industries as well. Hydrocyclones utilize centrifugal forces to accelerate the settling rate of particles. Since their mechanical struc-ture is simple, durable, and relatively inexpensive, hydrocyclones are one of the most popular mineral separation devices found in industry. They are used in closed-circuit grinding, de-sliming circuits, de-gritting procedures, and thick-ening operations. Figure 11.12 shows a schematic of a typical hydrocyclone. The lower part of the hydrocyclone has a conical vessel shape with an opening at the apex that allows the coarse or heavier particles to be removed. The top of the hydrocyclone is a cylindrical section that is closed with the exception of an overflow pipe called a vortex finder. This vortex finder prevents the mineral sample from going directly into the overflow while allowing the fine particles to remain in the hydrocyclone. The actual mineral separation occurs in this cylin-drical section due to the existence of a complex velocity distribution that carries the fine particles out the top and the coarse particles to the apex.

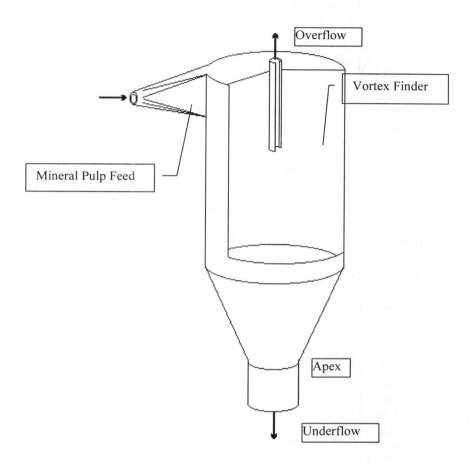

Figure 11.12 – A typical hydrocyclone.

There have been numerous efforts in the area of modeling hydrocyclones. Plitt [12] developed a statistical model to predict the split size of the hydrocyclone, D_{50}. The split size is that size particle that has a 50 percent chance of exiting in either the overflow or the underflow. The Plitt's model has proven to be quite robust and is often cited in the literature. Despite the numerous efforts to model hydrocyclones, none of the models has been universally adapted. Most of the models are either too computationally intensive or are only applicable to a limited range of hydrocyclone designs. To facilitate this hydrocyclone modeling effort, hydrocyclone performance data has been acquired from the U.S. Bureau of Mines. This data pertains to a wide range of designs used on a variety of mineral samples including iron, phosphate, and copper. A computer model of a hydrocyclone must consider several input parameters to compute a value of D_{50}. Figure 11.13 is a schematic of a hydrocyclone model where:

D_c = diameter of the hydrocyclone,
D_i = diameter of the slurry input,
D_o = diameter of the overflow,
D_u = diameter of the underflow,
h   = height of the hydrocyclone,
Q   = volumetric flow rate into the hydrocyclone,
ϕ = percent solids in the slurry input,
ρ = density of the solids,
D_{50} = hydrocyclone split size.

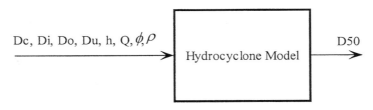

Figure 11.13 - Hydrocyclone computer model diagram.

The data available for the model development consists of input-output pairs. The inputs consist of the variables shown entering the model in Figure 11.13. The output is the D_{50}. The data will be segregated into two distinct groups: (1) a training set and (2) a test set. In training data set, the desired output data are provided to fuzzy system to update the parameters. The desired output is not given in the test set. NN with BORN was applied to model this hydrocyclone separation problem by the author of this chapter. However, due to the difference of system input and output ranges between training and test data sets, NN performance was found to be unacceptable. In order to adapt previously optimized NN by BORN, GA-BORN adaptive unit was used.

The NN that will be optimized by BORN has eight inputs, ten hidden neurons, and one output. This NN is trained with a learning rate of 0.01 for 75 epochs and then the BORN algorithm is executed for 25 epochs. The NN optimized by GA has the same structure as the one optimized using BORN. In order to optimize NN connections by GA, GA parameters are set to 30 populations, 19 strings, 0.8 probability of crossover, and 0.01 probability of mutation for 100 generations.

Figure 11.14 shows the training result for BORN. Despite the high non-linearity of hydrocyclone model, training performance for BORN was fairly good. Figures 11.15 to 11.19 show BORN test results before and after GA adaptive unit is executed. These plots indicated that the BORN algorithm was improved with the addition of a GA adaptive unit. However, it was not a large improvement. Figures 11.20 and 11.21 show the results of an NN in which the number of hidden neurons is increased to 50. Figures 11.22 and 11.23 are the results with 100 hidden neurons. The performance was improved dramatically because GA has more selections to adjust NN connectivities.

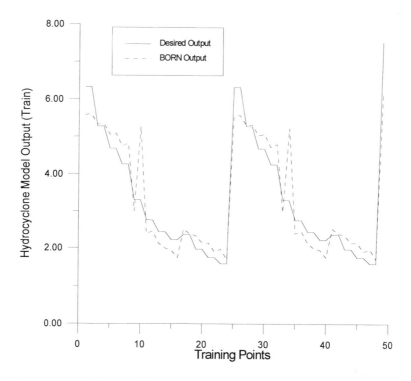

Figure 11.14 - BORN training result.

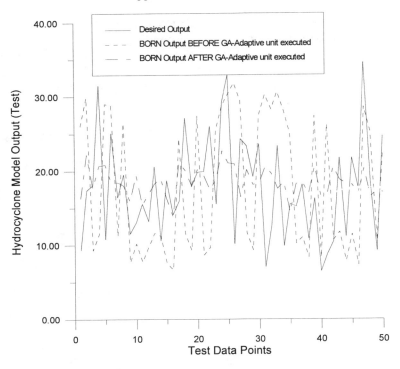

Figure 11.15 - BORN test result.

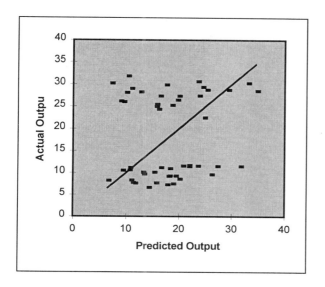

Figure 11.16 – Before GA-NN.

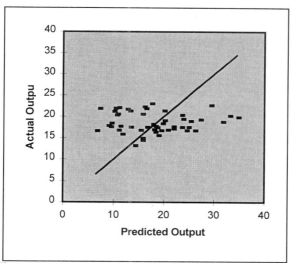

Figure 11.17 - After GA

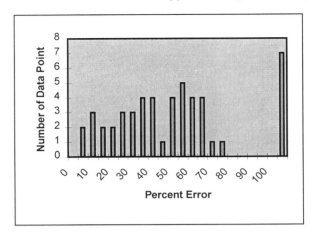

Figure 11.18 – Before GA.

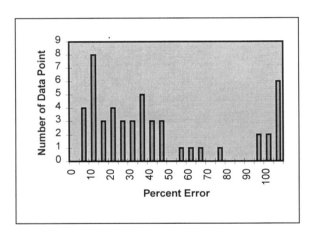

Figure 11.19 - After GA.

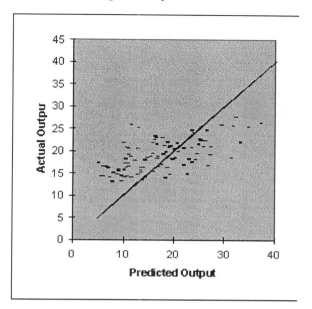

Figure 11.20 – GA-NN opt. (50 hidden).

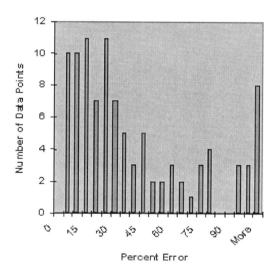

Figure 11.21 - GA-NN opt. (50 hidden).

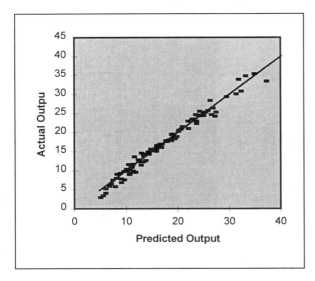

Figure 11.22 – GA-NN opt. (100 hidden).

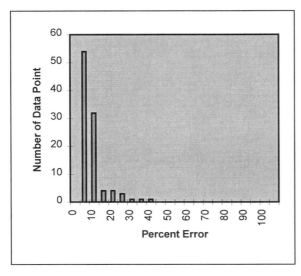

Figure 11.23 - GA-NN opt. (100 hidden).

CONCLUSION

GAs are highly efficient and robust optimization algorithms that have been used effectively in a variety of disciplines. In this chapter, simple GA was applied to enhance the performance of Bama Optimized Recurrent Neural Networks algorithm. In order to measure the performance of BORN with GA, two different studies are considered.

First, BORN was optimized with a GA and the result was compared with the result that was generated by only the BORN algorithm. Although GA-NN found a near optimal solution, the BORN algorithm performed better in a training case.

Second, a GA was implemented as a BORN connectivities adaptive unit to improve the BORN algorithm. The GA improved its performance for a hydrocyclone modeling problem. However, the GA needed a fairly large NN structure to minimize the error. The advantage of this method is that a GA can improve the performance of BORN by changing only the NN connections.

REFERENCES

[1]   Montana, D., J. and Davis, L. (1989) *Training feedforward neural networks using genetic algorithms*, Proceedings of 11th International Joint Conference on Artificial Intelligence, pp. 762-767, San Mateo, CA.

[2]   Goldberg, D. E., (1989) *Genetic algorithms in search, optimization, and machine learning.* Reading, MA: Addison Wesley.

[3]   Belew, R.K., McInerney, J., and Schraudolph, C. (1990) *N.N. Evolving networks: Using genetic algorithms with connectionist learning.* CSE technical report CS90-174, La Jolla, CA: University of California at San Diego, June.

[4]   Anderson, C. W. (1989) *Learning to control an inverted pendulum using neural networks*, IEEE Control Systems Magazine, 9, 31-37.

[5]   Wieland, A. P. (1990) *Evolving neural network controllers for unstable systems*, IEEE International Joint Conference on Neural Networks, pp. II-667 - II-672, Seattle, WA.

[6]   Werbos, P. J., (1992) *Neural Networks, System Identification, and Control in the Chemical Process Industries*, Handbook of Intelligent Control, Ed. White and Sofge, Van Nostrand Reinhold, New York, NY.

[7]   Narendra, K. S., (1992) *Adaptive Control of Dynamical Systems Using Neural Networks*, Handbook of Intelligent Control, Ed. White and Sofge, Van Nostrand Reinhold, New York, NY.

[8]   KrishnaKumar, K., (1993) *Optimization of the neural net connectivity pattern using a back-propagation algorithm,* Neurocomputing 5, p273-286.

[9]   KrishnaKumar, K. and Nishita, K., *Robustness of Recurrent Neural Networks*, WCNN'96, San Diego, CA, June 1996.

[10]   KrishnaKumar, K. and Nishita, K., (1995) *BORN – Bama Optimized Recurrent Neural Networks* WCNN'95, San Diego, CA.

[11]   Willis, B.A., (1979) *Mineral Processing Technology*, Toronto: Pergamon Press.

[12]   Plitt, L.R., (1976) *A mathematical model of the hydrocyclone classifier*, CIM Bulletin, 69, pp. 114-123.

CHAPTER 12

GAUSS-LEGENDRE INTEGRATION USING GENETIC ALGORITHMS

Barry Weck, Charles L. Karr, and L. Michael Freeman
Department of Aerospace Engineering and Mechanics
University of Alabama
Box 870280
Tuscaloosa, AL 35487-0280
e-mail: bweck@eng.ua.edu

ABSTRACT

Numerical integration (quadrature) is a tool used by engineers and scientists to approximate definite integrals that cannot be efficiently solved analytically. Quadrature methods rely on the selection of *quadrature nodes* (locations at which the integrand function is to be evaluated) and *weights* (values used in the approximation of the integral) to approximate integrals. In most quadrature methods, the weights are selected to give quality results while the quadrature nodes are distributed uniformly over the limits of the integral. In Gauss-Legendre quadrature, both the quadrature nodes and the weights are selected, thereby producing accurate results with minimal computation. There is, however, a problem with this approach: a system of nonlinear equations must be solved for the quadrature nodes and the weights. Solving systems of nonlinear equations is perhaps the most difficult problem in all of numerical computation. In this chapter, an approach to solving the system of nonlinear equations needed to define a Gauss-Legendre numerical integration is presented in which a genetic algorithm (GA) is used in conjunction with a traditional Newton method. Results indicate that the hybrid of a GA and Newton's method is effective and represents an efficient approach to solving the systems of nonlinear equations that arise in the implementation of Gauss-Legendre numerical integration.

GAUSS-LEGENDRE INTEGRATION

Numerical integration (also called *quadrature*) is a primary tool used by engineers and scientists to approximate definite integrals that cannot efficiently be solved analytically. The history of quadrature extends back to the beginnings of calculus and before. Thus, there are a number of accepted methods for numerically integrating a function [1]. However, with the advent of the automatic computer, the study of quadrature methods has lost some of its luster because it is not a terribly difficult numerical problem. However, one particular method of quadrature, Gauss-Legendre quadrature, proves to be especially accurate and efficient because it seeks to choose two parameter sets whereas other quadrature methods generally focus only on the selection of one parameter set. However, this robustness does not come without a cost: implementing Gauss-Legendre quadrature results in a system of nonlinear equations that must be solved for the two parameter sets.

The basic problem of quadrature is to approximate the definite integral of f(x) over the interval [a, b] by evaluating f(x) at a finite number of sample points. An example quadrature formula is:

$$Q[f] = \sum_{j=0}^{N} w_j \, f(x_j) = w_0 \, f(x_0) + w_1 \, f(x_1) + \ldots + w_N \, f(x_N) \qquad (12.1)$$

with the property that

$$\int_a^b f(x)\, dx = Q[f] + E[f]$$

where Q[f] is the approximation to the integral and E[f] is the truncation error associated with the approximation. The values x_j are called the *quadrature nodes* and the values w_j are called *weights*.

Most quadrature methods such as the rectangular rule, Simpson's rule, and Romberg integration assume that the quadrature nodes are equally spaced over the interval [a, b] as shown in Figure 12.1a below. Thus, these algorithms focus on the selection of weight values that result in accurate approximations of the integral. However, Gaussian quadrature methods attempt to select the quadrature nodes in conjunction with the weight values so that the approximation is as accurate and efficient as possible. This approach generally results in non-uniformly distributed quadrature nodes as shown in Figure 12.1b. Gauss-Legendre quadrature is a particular Gaussian method that proves to be especially robust – accurate and efficient.

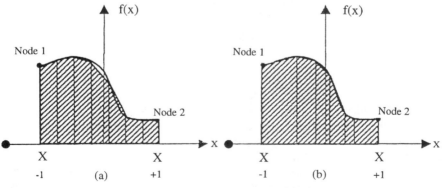

Figure 12.1 – (a) Most quadrature methods typically use a uniform distribution of the quadrature nodes, and focus on the selection of weights. (b) Gauss-Legendre quadrature simultaneously chooses values of both quadrature nodes and weights. Thus, the distribution of the quadrature nodes is rarely uniformly distributed.

Gauss-Legendre quadrature is used to determine the area under the curve $y = f(x)$ on the interval [-1, 1]. There are a number of popular versions of the method, and these variations are distinguished by the number of points (quadrature nodes) at which the function is to be evaluated. As an example of the rationale of computing the location of the quadrature nodes as is done in Gauss-Legendre integration, consider a most simple two-point formula. Consider a case in which a trapezoid rule is used (thus the two weights w_1 and w_2 are defined) to approximate the area under the curve defined by f(x). In a two-point formula, the function is to be evaluated at only two points. In a classic trapezoid rule, the assumption is that the quadrature nodes are equally spaced and thus, the function is evaluated at the two limits of integration -1 and 1. However, as can be seen in Figure 12.2, if the quadrature nodes are selected more carefully, the two-point trapezoidal formula yields a more accurate approximation to the integral.

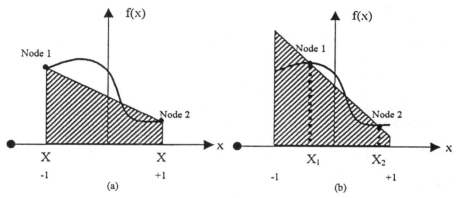

Figure 12.2 - (a) The two-point trapezoidal rule can be used to approximate the area under the function f(x) on the interval [-1, 1]. This form of quadrature assumes that the quadrature nodes are equally spaced. (b) A two-point trapezoidal rule in which the location of the quadrature nodes is computed can give a more accurate approximation than the traditional trapezoid rule.

A general Gauss-Legendre N-point formula can be developed that is exact for polynomial functions of degree $\leq (2N - 1)$ [2]. The general N-point formula results in a system of nonlinear equations that must be solved to compute the value of the quadrature nodes (x_j) and their associated weights (w_j). The system of nonlinear equations that result from the general N-point formula is:

$$f_n(w_1, w_2, \ldots, w_N, x_1, x_2, \ldots, x_N) = \sum_{j=1}^{N} w_i\, x_i^{n-1} - \int_{-1}^{1} x^{n-1} = 0$$

(12.2)

$$for\ n = 1, 2, \ldots, 2(N - 1),\ 2N$$

As an example, the system for the two-point formula (N = 2) results in the following system of four nonlinear equations:

$$w_1 + w_2 - 2 = 0$$

$$w_1\, x_1 + w_2\, x_2 - 0 = 0$$

(12.3)

$$w_1\, x_1^2 + w_2\, x_2^2 - \frac{2}{3} = 0$$

$$w_1\, x_1^3 + w_2\, x_2^3 - 0 = 0$$

This system of four nonlinear equations is difficult to solve using traditional search techniques such as a Newton method for two reasons. First, the values of x_1, x_2, w_1, and w_2 should be determined to several decimal places. For instance, it is not unusual to see these values computed to ten decimal places. Second, the ability of a method to converge to the solution in this problem is highly sensitive to the quality of the first guess supplied. This sensitivity to the initial guess will be analyzed in detail in a later section. However, to give the reader a feel for this sensitivity, consider the following exercise. A Newton method was supplied with random initial starting guesses in an attempt to solve for the parameters in a Gauss-Legendre four-point formula (x_1, x_2, x_3, x_4, w_1, w_2, w_3, and w_4). In 100,000 randomly supplied initial guesses, a Newton method converged to the correct solutions only 792 times – it **converged** only 792 times, not found the correct answer 792 times. As will be seen later, this problem grows exponentially worse as the parameters for higher order Gauss-Legendre quadrature routines are sought.

In this chapter, a hybrid scheme for solving the nonlinear system of equations resulting from Gauss-Legendre quadrature is discussed. The scheme involves using a genetic algorithm (GA) to locate initial guesses supplied to a traditional Newton search. The effectiveness of this approach is demonstrated on two, three, and four-point Gauss-Legendre formulae. It is important to note here that the goal in this chapter is to determine the **values of the coefficients** as accurately as possible, not to locate parameters that give close approximations to a particular integral. The latter is a much simpler search problem because the system of nonlinear equations does not necessarily have to be solved exactly to accurately compute a given integral. But, before the hybrid scheme is introduced, a brief overview of methods for solving systems of nonlinear equations is provided.

SYSTEMS OF NONLINEAR EQUATIONS

Solving systems of nonlinear equations is one of the most difficult problems in numerical computation. In fact, Press et al. [1] have gone so far as to state that "there are no good, general methods for solving systems of nonlinear equations." The difficulties associated with solving systems of nonlinear equations are magnified as the number of equations increases, but it is not unheard of for even five equations to be very difficult to solve. Since the problem can be exacting for even small systems, and since there are numerous engineering applications in which systems of nonlinear equations must be solved, this is an area in which potentially vast amounts of computational time can be saved. There are a myriad of engineering applications for these problems from such diverse areas as electric power generation and distribution, multi-objective optimization, and trajectory/path planning in which large systems of nonlinear equations must be solved efficiently. But, in this chapter, the focus will be on a specific family of systems of equations: those resulting from the implementation of Gauss-Legendre quadrature.

The problem of solving a system of nonlinear equations is to select a vector of solutions **x** such that a vector of functions **f** is driven simultaneously to zero. To gain some insight into the problem, consider a most simple example in which the goal is to simultaneously drive two functions $f(x, y)$ and $g(x, y)$ to zero:

$$f(x, y) = 0$$

$$(12.4)$$

$$g(x, y) = 0$$

The two functions f and g are arbitrary functions, each of which has zero-contour lines that divide the x-y plane. Points are sought at which the zero-contour lines of the two functions intersect. To help gain an appreciation for the difficulty of this problem, consider the zero-contour lines of two sample functions shown below in Figure 12.3. Here, the zero-contour lines of the two functions intersect at four points (M_1 through M_4). Thus, there are four (x, y) pairs for which both functions are simultaneously driven to zero.

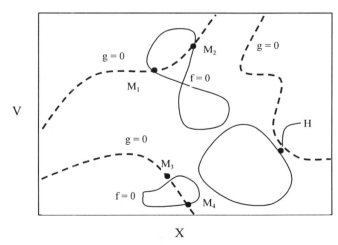

Figure 12.3 - The solution to the problem of interest occurs at points M_1 - M_4 where the zero-contour lines intersect. Point H misleads many methods because the zero-contour lines approach one another yet do not intersect.

There are at least two difficulties associated with the two-dimensional problem represented by Figure 12.3. First, there are four solutions to the problem. Often in nonlinear systems of equations, there are multiple roots. Second, and generally more difficult to overcome, the functions f and g are not necessarily related to one another. There is nothing special about the common points of zero-contour lines from either f's or g's perspective. Thus, in order to solve this problem completely, the entire zero-contour lines of each function involved must be mapped out. Furthermore, this is an example of only two functions of two independent variables. When the system of equations is expanded to N-dimensions, the goal is to determine the intersection of N unrelated zero-contour hyperplanes, each of dimension (N-1). Fortunately, there are alternatives to the exhaustive mapping of zero-contour lines. A few of the methods commonly used for determining the roots of single equations are extensible to higher dimensions.

There are a number of effective methods for solving one nonlinear equation. Of the most popular of these, bisection [3], regula falsi [1], and modified regula falsi [1] cannot be extended to systems of nonlinear equations because they depend on patterns of sign changes in the function being driven to zero. These patterns of sign changes do not uniquely identify the location of the zeros of several functions of several variables. The fixed point iteration method [4] is perhaps the simplest of root-finding algorithms and is easily extended to systems of equations. However, the fixed point method is highly susceptible to divergence in one dimension, and the method diverges too frequently in systems of equations to be practical. Both Muller's method [3] and higher-order Newton methods extend in principle, but the computational effort associated with each of these methods increases rapidly with the number of equations being solved. Thus, these methods are rarely used. The principal method for solving a system of nonlinear equations is the traditional Newton method [2], a derivative-based search, because of its simplicity and efficiency. However, since Newton's method is driven by derivative information, its performance and convergence characteristics are highly dependent on the initial guess of the solution with which it begins.

Of course, there are alternatives to extending the methods of root-finding in a single dimension. One popular alternative to the one-dimensional root-finding algorithms is to collapse the higher dimensionality of the problem into a single dimension by adding the sums of squares (or the absolute values) of the individual functions f_i to get a master function F that is to be minimized where

$$F(x_1, x_2, \ldots, x_N) = \sum_{j=1}^{N} f_i^2(x_1, x_2, \ldots, x_N) \qquad (12.5)$$

This master function is positive definite and has a global minimum at each of the roots to the equation, M_i. However, the function is often fraught with local minima that tend to exist at points where the zero-contour lines of two functions approach but do not overlap one another. Such a point is labeled H in Figure 12.3.

Unfortunately, another of the most difficult problems in numerical methods is the minimization of functions with numerous local minima.

There are a number of efficient algorithms for minimizing a function of many variables. However, the majority of these methods are not always effective in the minimization of the master function stemming from the problem of solving a system of nonlinear equations. Most are incapable of locating multiple roots, or they tend to converge to local minima. Optimization techniques such as variable metric methods [5], Brent's method [6], and simplex search [7] prove inadequate in the general N-dimensional nonlinear equation problem.

Genetic algorithms consider multiple solutions to search problems simultaneously due to their population approach. Thus, they are effective in optimization problems with more than one optimum solution [8]. Further, since genetic algorithms do not use derivative information, they do not tend to get caught in local optima. The above two attributes of GAs allow them to overcome some of the shortcomings that prevent more traditional search techniques from effectively solving systems of nonlinear equations. However, genetic algorithms do not always converge to the true minimum in a search problem. Thus, they are not inviting stand-alone tools for minimizing the master function $F(\mathbf{x})$ resulting from the problem of solving a system of nonlinear equations. The strength of genetic algorithms is that they rapidly converge to *near-optimal* solutions.

In this chapter, a hybrid approach to solving the systems of nonlinear equations resulting from the application of Gauss-Legendre quadrature is presented. The higher dimensionality of the problem is collapsed into a single dimension by adding the absolute values of the individual functions f_i, thereby forming a master function $F(\mathbf{x})$ that is to be minimized. A GA is used to rapidly converge to near-optimal solutions of the system of equations. These near-optimal points are then used as the initial guesses employed by a traditional Newton search.

A HYBRID APPROACH

A Newton-Raphson algorithm lies at the heart of the hybrid approach presented here. This method suffers from its sensitivity to the initial guess it is supplied, but it is still the most popular algorithm for solving systems of nonlinear equations. Once the neighborhood of a root has been identified, Newton's method provides an efficient means of converging to the root, if it exists, or of spectacularly failing to converge, indicating that there is no root in the neighborhood. Thus, the focus of the hybrid scheme is to effectively locate regions in which roots are likely to exist; a problem that proves to be readily solved using a GA.

Newton's method is driven by derivative information and is derived from Taylor's series analysis. In the solution of systems of equations, the method employs a linear model to solve the equations. A linear model of $\mathbf{f}(\mathbf{x})$ is of the form

$$f(x) = a_0 + J * (x - x_0)$$

(12.6)

where J is a matrix. Usually $a_0 = f(x_0)$, and Equation (12.6) is then the first two terms of the Taylor's series of $f(x)$ if the model J uses the first derivatives. We want to have $f(x) = 0$, and so set the model in Equation (12.6) equal to zero and solve for x to obtain

$$x = x_0 - J^{-1} f(x_0)$$

(12.7)

An iteration is formed from Equation (12.7) above in an obvious way. Note that an N by N system of linear equations must be solved at each iteration. Thus, such derivative-based methods can be time consuming.

Newton's method for systems of nonlinear equations uses the linear model based on the "tangent planes." Since $f(x) = 0$ is N equations

$$f_i(x) = 0 \qquad i = 1, 2, \ldots, N$$

$y = f_i(x)$ defines a surface. At x_0, the tangent plane to the surface is

$$f_i(x_0) + \sum_{j=1}^{N} \frac{\partial f_i}{\partial x_j}(x_j - x_{0j})$$

The matrix J is the *Jacobian* matrix

$$J = \begin{vmatrix} \dfrac{\partial f_1}{\partial x_1} & \dfrac{\partial f_1}{\partial x_2} & \cdots & \dfrac{\partial f_1}{\partial x_N} \\[2mm] \dfrac{\partial f_2}{\partial x_1} & \dfrac{\partial f_2}{\partial x_2} & \cdots & \dfrac{\partial f_2}{\partial x_N} \\[2mm] \cdots & \cdots & \cdots & \cdots \\[2mm] \dfrac{\partial f_N}{\partial x_1} & \dfrac{\partial f_N}{\partial x_2} & \cdots & \dfrac{\partial f_N}{\partial x_N} \end{vmatrix}$$

(12.8)

The linearized model employed by Newton's method is quite effective if the initial guess of x_0 is reasonably close to a root. However, the development of the coefficients required in Gauss-Legendre quadrature is such a case; after all, if the values of the coefficients are approximately known, then the value of the integral can more than likely be accurately approxmiated. Thus, in the hybrid approach proposed, a GA is employed to locate regions in which roots to the system of equations are likely to exist.

GAs are search algorithms based on the mechanics of natural genetics. Despite the numerous variations to Holland's basic GA [9], the majority of evolutionary algorithms proceed as follows:

> 1. generate a population of strings coded to represent potential solutions to the search problem at hand;

> 2. evaluate the effectiveness or quality of the solutions represented by the strings in the current population (compute a *fitness function* value for each string);

> 3. make copies of strings in proportion to the effectiveness of the solutions they represent (in proportion to their fitness function values) and combine portions of these highly fit strings according to genetic operators to produce a new population of strings (a new *generation*);

> 4. repeat steps 2 and 3 until a suitable solution to the search problem has been found.

The different varieties of evolutionary algorithms occur due to the genetic operators, the definition of the coding scheme, and the fitness function. In this chapter, the genetic operators are from the *simple genetic algorithm* due to Goldberg [8]. Computer code for implementing the algorithm can be found in the reference by Goldberg [8].

The coding scheme for this problem is accomplished using a relatively standard coding scheme. A *concatenated, binary, linearly mapped coding* scheme is employed [8]. In this scheme, the parameters of the search problem are represented as bit strings. For example, in the two-point Gauss-Legendre problem, there exist four equations in four unknowns. The equations are presented collectively as Equation (12.3) and the unknowns are w_1, w_2, x_1, and x_2. The four parameters are each coded as binary strings according to the formula

$$p = p_{min} + \frac{b}{(2^l - 1)}(p_{max} - p_{min})$$
(12.9)

where p is the parameter being coded, p_{min} is the minimum value of the parameter (in the case of x_i, the minimum value is -1, while for w_i, it is 0), p_{max} is the maximum value of the parameter (in the case of both x_i and w_i, the maximum values are 1), and b is the decimal value associated with a binary string of length l.

Each of the parameters in the search problem is represented in this fashion, and then concatenated to form a single string that represents a complete solution to the problem. To achieve the accuracy required in this application, bit strings of length forty were employed.

The fitness function is based on identifying the zero-contours of the individual functions f_i. The fitness function to be minimized by the genetic algorithm is

$$g = \max \ [abs \ (\ f_i \ )] \qquad for \ i = 1, 2, ..., N \qquad (12.10)$$

where g is the fitness function to be minimized, $max[abs(f_i)]$ is the maximum value of the absolute value of individual equations in the system $\mathbf{f(x)} = 0$, and N is the number of equations in the system. This function g is a positive definite function that contains global minimums at each of the roots to the system of equations. Recall that a GA will not guarantee convergence to the roots of the equation, but it will rapidly locate regions in which the fitness function is minimal. For the problem of solving systems of nonlinear equations, these regions will likely contain roots to the equations or points at which the zero-contour lines approach but do not cross (false roots).

The hybrid approach is built around two concepts. First, once regions in which roots are likely to exist have been determined, a Newton method can quickly distinguish between "real roots" and "false roots." If there are roots in the region, a Newton method will rapidly converge to the roots. Second, a GA can be used to rapidly determine regions in which roots are likely to exist. It does so by minimizing a fitness function that indicates promising regions in the search space. The solutions determined by the GA are used as initial guesses supplied to a Newton-Raphson method. As will be seen in the next section, the result is a hybrid scheme that is far more efficient than a Newton method alone.

RESULTS

The hybrid scheme described in the previous section has been used to solve the systems of nonlinear equations that arise in the development of N-point Gauss-Legendre quadrature formulae. The goal here is to solve the system of nonlinear equations, not to determine values of quadrature nodes and their associated weights that allow for accurate integral approximations. Thus, results will focus on the values of the quadrature nodes and the weights, not on the accuracy of an approximation to any given integral. The quadrature nodes and their associated weights for various Gauss-Legendre quadrature formulae resulting from the solution of the system of equations depicted in Equation (12.2) are shown in Table 12.1. Notice that the given values are accurate to the tenth decimal place [2].

Table 12.1 - The quadrature nodes and the associated weights for four-, five-, six-, seven-, and eight-point Gauss-Legendre quadrature formulae are provided below.

	Quadrature nodes, x_i	weights, w_I
2	-0.5773502692 0.5773502692	1.0000000000 1.0000000000
3	±0.7745966692 0.0000000000	0.5555555556 0.8888888888
4	±0.8611363116 ±0.3399810436	0.3478548451 0.6521451549
5	±0.9061798459 ±0.5384693101 0.0000000000	0.2369268851 0.4786286705 0.5688888888
6	±0.9324695142 ±0.6612093865 ±0.2386191861	0.1713244924 0.3607615730 0.4679139346
7	±0.9491079123 ±0.7415311856 ±0.4058451514 0.0000000000	0.1294849662 0.2797053915 0.3818300505 0.4179591837
8	±0.9602898565 ±0.7966664774 ±0.5255324099 ±0.1834346425	0.1012285363 0.2223810345 0.3137066459 0.3626837834

As a point of comparison for gauging the effectiveness of the hybrid scheme, a Newton method was provided with randomly generated initial guesses for the systems of equations. The results of this exercise are shown in Figures 12.4 and 12.5. Figure 12.4 shows the number of successes in 100,000 randomly generated guesses. Here, a "success" means convergence to the correct solution to the system of equations. Figure 12.5 shows the number of random initial guesses required before the initial success. Thus, when determining the coefficients for the seven-point formula (which involves solving a system of fourteen nonlinear equations), the Newton method converged only 4 times for the 100,000 initial guesses provided. The expected success rate of randomly guessing is 4/100,000 = $4*10^{-5}$.

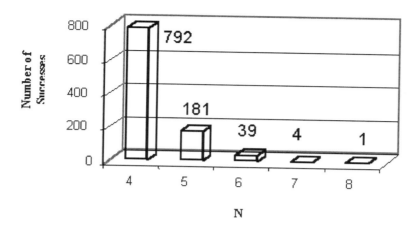

Figure 12.4 - As can be seen above, the number of "successes" in 100,000 randomly generated initial guesses decreases exponentially with N for the Newton method.

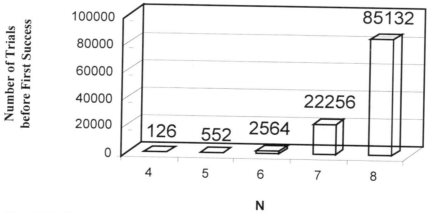

Figure 12.5 – The number of guesses required before the initial successful convergence by the Newton method increases exponentially with N for the Newton method.

So that statistics suitable for comparison could be computed, the hybrid scheme was implemented in the following way. In the course of the GA's operation, the best solution existing at the end of each generation was sent as an initial guess to a Newton method. The Newton method used this initial guess in an attempt to solve the system of equations. The results are shown in Figure 12.6. When considering this figure, recall that a fitness function evaluation consisted only of computing the numerical value of 2N individual equations and taking the maximum of these values. Thus, a fitness function evaluation is actually quite inexpensive computationally. The number of fitness function evaluations shown in Figure 12.6 are the number of evaluations required such that the Newton method would converge to the solution of the system of nonlinear equations on this first guess. Notice in this figure that the number of function evaluations required by the GA increases polynomially with N as opposed to the exponential behavior depicted by the Newton method acting independently.

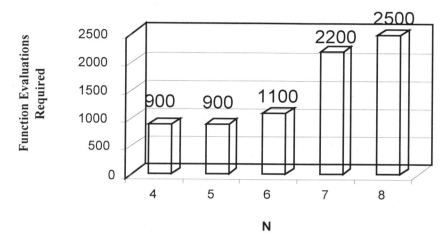

Figure 12.6 – The success rate of the hybrid scheme shows that the number of function evaluations increases polynomially as opposed to the exponential increase required by the Newton method.

To summarize the results presented in this section: when the hybrid scheme is used, a GA can effectively locate an initial guess that allows the Newton method to converge to an accurate solution much more quickly than when random initial guesses are supplied. It is also important to note that there is some computational overhead associated with the operation of the GA. The standard genetic operators require the generation of random numbers and the combining and copying of bit strings. Additionally, the fitness function values must be computed. However, this computational overhead is negligible when compared to the computation involved with computing the inverse of the Jacobian matrix.

SUMMARY

A hybrid scheme for solving the systems of nonlinear equations resulting from the application of Gauss-Legendre quadrature has been described and tested. Gauss-Legendre quadrature is an effective numerical integration technique because it simultaneously selects the best values of both the quadrature nodes and the weights, whereas most quadrature methods focus solely on the value of the weights. However, solving the systems of nonlinear equations resulting from the implementation of this quadrature method is difficult. Thus, a hybrid scheme capturing the strengths of both a traditional Newton method and a GA was developed. The hybrid scheme combines the rapid convergence characteristics of the derivative-based Newton-Raphson method (once a quality initial guess is determined) and the global search capabilities of GAs. The result is a method that dramatically reduces the computational time and effort required to implement Gauss-Legendre quadrature.

REFERENCES

[1]   Press, W. J., Flannery, B. P., Teukolsky, S. A., & Vetterling, W. T. (1988). *Numerical recipes in C: The art of scientific computing.* Cambridge: Cambridge University Press.

[2]   Mathews, J. H. (1992). *Numerical methods for mathematics, science, and engineering.* Englewood Cliffs, NJ: Prentice Hall.

[3]   James, M. L., Smith, G. M., & Wolford, J. C. (1977). *Applied numerical methods for digital computation with FORTRAN and CSMP, 2nd edition.* New York, NY: Harper & Row Publishers.

[4]   Rice, J. R. (1983). *Numerical methods, software, and analysis.* New York, NY: McGraw-Hill Book Company.

[5]   Polak, E. (1971). *Computational methods in optimization.* New York, NY: Academic Press.

[6]   Acton, F. S. (1970). *Numerical methods that work.* New York, NY: Harper and Row Publishers.

[7]   Nelder, J. A. & Mead, R. (1965). A simplex method for function minimization. *Computer Journal*, 7, 308-321.

[8]   Goldberg, D. E. (1989). *Genetic algorithms in search, optimization, and machine learning.* Reading, MA: Addison-Wesley.

[9]   Holland, J. H. (1975). *Adaptation in natural and artificial systems.* Ann Arbor, MI: The University of Michigan Press.

CHAPTER 13

USING GENETIC OPERATORS TO DISTINGUISH CHAOTIC BEHAVIOR FROM NOISE IN A TIME SERIES

John Nilson
AT&T Corporation
600 Mountain Avenue
Murray Hill, NJ 07974
email: jvn@insight.att.com

ABSTRACT

In this study, genetic operators are used in reproducing an important study that distinguishes chaotic behavior from noise in a time series. Distinguishing chaotic behavior from noise is accomplished by predicting future time values from current time values. With chaotic behavior, the prediction accuracy declines as the time horizon increases while with noise, the accuracy essentially remains the same. The results are important for analysts in order to apply appropriate methodologies to nonlinear dynamics. It is especially useful to forecasters who use time series as part of their methodology.

INTRODUCTION

According to G. Sugihara and R. May, authors of an important article entitled: *Nonlinear forecasting as a way of distinguishing chaos from measurement error in time series* (Nature, vol. 344, April 19, 1990, pages 734-741), there are two sources of uncertainty in forecasting the behavior of dynamical systems: 1) the error associated with measurement (e.g., sample size error, unpredictable environments), which is often called noise, and 2) the complexity of the system itself, which often exhibits chaotic behavior. Both of these sources of uncertainty appear as randomness in time series.

If an analyst could understand which of the two sources of uncertainty he/she was dealing with, he/she could take appropriate steps to improve the analysis. For example, if the source were identified as noise, then the analyst might choose a smoothing or a filtering technique as part of the methodology. If, however, the source were due to the inherent nature of the system, then the analyst might approach the analysis in a different manner. He/she might use nonlinear techniques such as neural networks, fuzzy expert systems, or genetic algorithms as the basis of their methodology. Thus the ability to distinguish between the two sources can be important in the approach and in the methodology of the analysis.

0-8493-9801-0/98/$0.00+$.50

Some attempts have been made in the past to distinguish the two sources (8) but these have had mixed results and require large sample sizes (10,000 to 20,000 sample points). The study by Sugihara and May uses a non-parametric approach in order to reduce the required number of required sample points. A standard technique, called Time Delay Embedding, is used to create a geometric space using the time series values as components. The "smallest" simplex is then formed by using the nearest neighbors (of the current point in the series) as vertices. The simplex is projected into the future by keeping track of where the vertices end up after n number of time steps. By applying the weights established in the original simplex (calculated using the current point in relation to the vertices) to the projected simplex, they obtain a forecast of the current point.

The study is important because, as a non-parametric technique, there are no underlying assumptions about distributions or population parameters. It allows for the central thesis of the study: that forecasts of time series with measurement error exhibit a high level of noise even as the forecast time horizon increases. The accuracy of forecasts then tend to be constant for all time horizons. Forecasts of time series with chaotic behavior, on the other hand, get worse as the time horizon increases. The accuracy of the forecasts then tends to fall off as the time horizon increases.

PROBLEM

In the study, an extensive search is made in determining the "smallest" simplex surrounding the current point. This involves a lengthy search for the exact set of points surrounding the current point comprising the "smallest" simplex. A key parameter for this is the embedding dimension. A simplex in an embedding dimension of three requires four points (four vertices), while a simplex in an embedding dimension of five requires six points (six vertices). Determining the exact set of four points (in an embedding dimension of three) out of say 1000 points requires an exhaustive number of searches using combinations of 1000 points, taken four at a time. For larger series, say, 10,000 points the cost increases at an exponential rate. (The combination of 10,000 points taken four at a time is more than ten times the combination of 1000 points taken four at a time.)

The approach presented uses genetic reproduction operators to replace the determination of the "smallest" simplex. Crossover and mutation operators are used after the projected simplex has been determined. It is anticipated that the application of genetic operators will duplicate the results of the original study, while eliminating the extensive search costs.

SOLUTION FORMATION

The following paragraph details the approach employed by Sugihara and May, called the Simplex Projection Method. The next section then demonstrates how genetic operators fit into this simplex projection methodology.

The basic idea of the Simplex Projection Method compares forecasts (using a specific time step) of future values with actual values in a time series. When a complete series is finished, the overall accuracy of the many individual forecasts is determined by use of the correlation coefficient. Then a higher time step is made, and the complete forecasting procedure is rerun. This process continues for time steps to a value of about 12 steps.

The results are then plotted. In a chaotic time series, the accuracy of the forecast diminishes as the horizon step increases; but with noise, the forecast error remains the same statistically and thus, the accuracy is somewhat constant as the horizon step increases. In their words, "comparing the predicted and actual trajectories, we can make tentative distinctions between dynamical chaos and measurement error: for a chaotic time series, the accuracy of the nonlinear forecast falls off with increasing prediction time interval, whereas for uncorrelated noise, the forecasting accuracy is roughly independent of prediction time interval [2]."

In order to make the comparison, a geometric space is generated from the time series values. This is a common methodology used in other nonlinear time series approaches such as neural networks. After obtaining a time series, they

1. Determine four very important parameters. These are:

    a. the time delay, tau - this is the number which is used for extracting values in the series that will become geometric components in a state space.

    b. the embedding dimension, E - is an estimate of the dimension of the underlying nonlinear structure from which the time series was obtained. These are often fractal in nature, meaning the dimension is a fraction, not a whole number. The embedding dimension is often the nearest integer of the fractal dimension.

    c. the size of the pattern library and the size of the test series. The pattern library is used to form the space, and the test series is used as an independent out-of-sample series for distinguishing noise and chaos.

    d. the time step - the number of time elements in the future that are being projected. The initial value is one and is incremeneted by 1 for a total of 12 time steps.

2. Establish a library pattern and a test series, using the parameters from 1 above. This is accomplished by converting the time series into a lagged delay coordinate space. The space (called a state space) consists of vectors determined from the parameters as follows: $x = [x_t, x_{t-tau}, x_{t-2\ tau}, \cdots x_{t-(E-1)\ tau}]$ where x's are individual time series elements. Each lagged coordinate vector represents a data point in an E-dimensional space.

An example of establishing data points in a state space from a time series is the following. Consider the small time series 12, 2, 8, 4, 11, 17, 19, 13, 17, 4, 9, 7, 16, 12, and 9. Assume a time lag of one, an "embedding dimension" of three, a pattern library size of six, and a test series of four. Then the following data points comprise the pattern library: (8,2,12), (4,8,2), (11,4,8), (17,11,4), (19,17,11), and (13,19,17). The following four vectors comprise the test series: (17,13,19), (4,17,13), (9,4,17), and (7,9,4).

3. Here is the expensive search. Take each data vector in the test series, one at a time, and compare it to each data vector in the pattern library and note the Euclidian distance between them. Obtain the E+1 closest vectors. Check to see if the four points form a simplex around the test vector. If not, obtain the next closest pattern vector and check, in ascending order of distance, if any four of the five points form a simplex around the test vector. If not, obtain the next closest pattern vector and check, in ascending order, if any four of the six points form a simplex. In short, the calculations involve an expanding (in combinations) set of data vectors that generate enormous search times as the size of the time series increases. Note if any of the data vectors form a simplex around the test vector, continue the process until the pattern library is exhausted. From the list of all simplices that were obtained, find the one with the smallest diameter.

4. Calculate the weighted distance from the test vector to each vertex (pattern vectors) for the selected simplex.

5. Project the simplex into the future using a selected time step, s, and then note where the points end up after s elements into the future. As previously stated, the reason for this assumption is that the projection falls off in a chaotic system, but, in the presence of noise, the projection is expected to be fairly constant. The measurement used in the projection (to indicate whether the projections fall off or not), is the correlation coefficient.

The following illustrates the projection using the example already discussed. Assume that the vector (17, 13, 19) represents the test vector and (13, 19, 17), (19, 17, 11), (17, 11, 4), and (11, 4, 8) represent the vertices of the simplex obtained in step three. Then the projection for the vertex at (13, 19, 17) is (17, 13, 19). Because the x component, i.e., the 13 projects to 17 (which is in fact the next element in the original time series), the x component, i.e., the 19, projects to 13, and the x component, a 17, projects to 19. Likewise, the vector at (7, 9,

4) projects to (16, 7, 9). For a time step value of 2, the projections for the vector (13, 19, 17) is (4, 17, 13) because the x component, a 13, skips to 3 forward elements in the original time series for a value of 4. The x component, a 19, projects two time elements for a value of 17, and the x component projects two time steps for a value of 13.

6. A new projected simplex is formed using the coordinates of the projected vertices.

7. Calculate the projected value of the test vector by using the weights previously established; calculate it as the weighted distance from the vertices of the projected simplex.

8. Record the difference between the actual results and the projected estimate. Then repeat the above process for each of the vectors in the test series. After all of the vectors in the test series have been used, calculate the correlation coefficient.

9. Repeat the above process for each time step. The time step starts at one and is incremented by 1 each cycle for a total of 12 cycles.

GENETIC OPERATIONS

The operations of reproduction, crossover, and mutation are used in applying genetic: concepts to the original study. It should be emphasized that it is small changes in the forecast that are important. These represent fluctuations that are present in the time series.

Reproduction and the Fitness Function

The fitness function replaces the simplex projection previously described. In this projection, the current point (a vector with the first coordinate as the current time value in a rolling series) associated with n nearby points. (In the study, this was necessarily a simplex requiring the extensive search.) The number n is the embedding dimension + 1, emulating the number of vertices in the simplex. These points are chosen as the nearest n neighbors (determined by Euclidean distance). The current position relative to the n points is calculated as the weighted distance to the points and saved for later use. The use of this type of function eliminates the extensive search to find the "smallest" simplex. The only difference between the weights using the simplex projection method and using genetic operations is that in the former case, all of the weights are positive while in the latter case, some of the weights can be (and probably will be) negative. As in the original study, we keep track of where the points end up after x forward time steps. From these, the forecast of the current point is made using the weights previously saved. Thus for every current time value in a rolling time series, one copy is created based upon the weighted distances from the nearest neighbors of the current point end up.

   To demonstrate the operation of the fitness function, it may be helpful to review the example from step 5 in the section entitled: Solution Formulation. In that example, the following vectors, in terms of components, were obtained directly from the time series:

Test vector (T) = (17,13,19)
Simplex Vertex 1 (V1) = (13,19,17)

These vectors were determined:

Vertex 2 (V2) = (19,17,11)	as the four nearest neighbors
Vertex 3 (V3) = (17,11,4)	to the test vector.
Vertex 4 (V4) = (11,4,8).	

Projected Simplex Vertex 1 (P1) = (17,13,19)
These were determined:

Vertex 2 (P2) = (19,17,11)	from the simplex above and
Vertex 3 (P3) = (17,11,4)	tracking s time steps from each
Vertex 4 (P4) = (11,4,8)	component in the series.

Goal: Use a fitness function to project the test vector s time steps into the future by:

1. Determining the relationship of the test vector to the original simplex. This is accomplished by writing the test vector (T) as a linear combination of the simplex vertices (V1,V2,V3,V4) to form:

$$T = w1*V1 + w2*V2 + w3*V3 + w4*V4, \tag{13.1}$$

where the w's are weights.

   The above system consists of three equations in four unknowns. An added restriction requires $w1 + w2 + w3 + w4 = 1$ to establish a relationship in a convex structure (in this case a tetrahedron). When the w's are either all positive or all negative, the test vector is encapsulated within the structure (simplex).

2. Using the fitness function to project the test vector into the future. By allowing the fitness function to operate without the strict requirement that all w's be either positive or negative, the exhaustive search to encapsulate the test vector is eliminated. The fitness function is then:

$$P = w1*P1 + w2*P2 + w3*P3 + w4*P4. \tag{13.2}$$

Assuming, for example, that the solution to finding values for the w's are: w1=.6, w2=.3, w3=.5, and w4=-.4, then the fitness function which projects the test vector s time steps into the future is given by:

$$T=.6*(17,13,19) + .3*(13,19,17) + .5*(19,17,11) -.4*(17,11,4).$$

Crossover

A value of about 0.6 will initially be used for the value of the probability of crossover. Multipoint crossover will be determined by choosing n number of random numbers of low order bits for each component value in the projected current value. Using only low order bit positions of the components is consistent with time series that are highly correlated, such as the stock market prices and other economic series.

Changes in prices are normally very small and reflect fluctuations at or near the decimal point.

Mutation

The probability of mutation will be initially set to about .001. For a sample size of 1000, only one operation of mutation is expected.

Coding Scheme

The coding scheme comes directly from the vector point, where the components are time series values. As previously described, each point will have the form $[x_t, x_{t-1}, x_{t-2}]$, assuming a dimension of three and time difference of one. Sixteen bits will be used as the size for each component in order to provide for a large variety of series. Thus, in the case of dimension 3 and time value 1, three components, each of 16 bits would be required to effectively encode the point. If we were interested in analyzing a time series such as the time series associated with the stock market (whose value is commonly about 6000.00 using the DJIA, including decimals), all 16 bits of each component would be used. An important reason to use this scheme is that for highly correlated time series (which are very common), the most recent values are the most important values. The next most important value is the previous value. Thus, the structure provides for the most effective case for genetic operations - keeping the most important allele adjacent to each other.

Using the example previously described, the coding scheme for the test vector of (17,13,19) would be the 48 bits as follows:

[0000 0000 0001 0001 0000 0000 0000 1101 0000 0000 0001 0011].

In a correlated time series, fluctuations in the values are represented as small changes in the low order positions of each vector. In this case, the low order bits would be bits 15,16,31,32,47, and 48. These low order bits would then be subject to crossover and mutation operations. Restricting the effects of cross over and mutation to the low order bits results in short, low order schemata.

PRODUCTION RUNS

The same two time series used by Sugihara and May were used in this study. The two time series were:

1. a tent map with parameter value 2, which is known to generate a chaotic time series [x_{t+1} = $2x_t$ for $0 < x < .5$, x_{t+1} = $2 - 2x_t$ for $.5 < x < 1$].

2. a modified sine map with 50% noise [x_t= $\sin(.5t) + N(-.5,.5)$].

For each of the above maps, about 1000 time series values were generated, representing 500 pattern and 500 test points. Vectors for both of these were created using the time delay methodology described in step 2 of the section entitled: Solution Formulation. Twelve runs were made for each map, each run representing one of the twelve time steps under review. Each run produced about 500 pairs of numbers: one for the result of the fitness function (projecting the test vector into the future) and one for the true value (obtained directly from the time series). The correlation coefficient was then calculated for each run (of about 500 pairs). The results are plotted in Chart 1, with similar results from the original study by Sugihara and May plotted in Chart 2. (Note: the horizontal axis of each chart represents an increasing number of time steps, while the vertical axis represents the correlation coefficient.)

RESULTS

Comparison of the two charts demonstrates that the application of genetic operators and the fitness function were successful in distinguishing chaotic behavior from noise. Both charts contain two graphs which behave in similar ways:

> a. they both contain a graph (Series 1) that is fairly constant in the horizontal direction. This graph represents the results of using noisy data (i.e., the sine map with 50% noise). The reason for the fairly constant graph is that the noise is independent of the time horizon. That is, it basically has the same value across the time different time steps.

> b. they both contain a graph (Series 2) that initially has a relatively higher correlation coefficient that decreases as the time horizon increases. This graph represents the results of using chaotic data (i.e., the tent map). The fall off of accuracy (in this case the correlation coeffi-

cient) is characteristic of chaotic systems. The technical term for this drop in accuracy is called "sensitive dependence on initial conditions" and means that two trajectories that are initially close become very distant over time. In the runs made, the two trajectories were represented by the projected test reults (from the fitness functions) and the actual results (obtained directly from the time series). When the two trajectories were close (at time step 1), the results showed a relatively higher correlation. As the time steps grew larger, the two trajectories separated more, resulting in a lower correlation between them.

Thus, the implementation of genetic operations was successful in obtaining the same characteristic results as obtained by Sugihara and May.

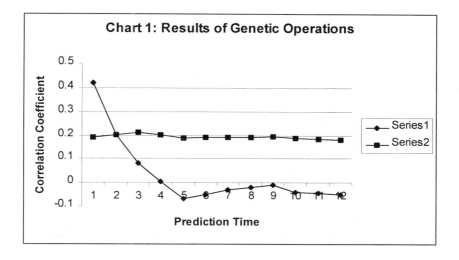

For comparison, the results obtained by Sugihara and May are plotted in Chart 2.

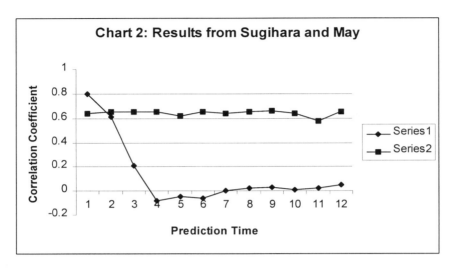

There is, however, one important difference between the two charts. The results obtained by Sugihara and May show significantly higher levels of correlation coefficients for both maps than do the results using genetic operators. Several attempts to improve the correlation coefficients using genetic operators were made, but no appreciable improvements were obtained. These attempts included changing the parameter values of the probabilities of crossover and mutation, varying the crossover points, experimenting with numerous new operators, and increasing the embedding dimension values. In particular, experiments using a shorter coding scheme did not appear to improve the level of the correlation coefficient.

CONCLUSION

Although numerous attempts were made to improve the correlation, the results may suggest that the benefits in using genetic operators to distinguish chaotic behavior from noise (i.e., significantly reduced search time) may come at a cost - accuracy. In some instances, this may not be acceptable. For example, when deciding to use filtering techniques, fuzzy expert systems, genetic algorithms, or some other approach to the analysis, the analyst might feel more comfortable with a higher level of accuracy. However, in other instances, it may be quite acceptable to work with a lower level of correlation. This could happen if the analyst wanted to obtain a quick understanding of the behavior of the underlying system.

REFERENCES

[1] Sugihara, G. & May, R. (1990). Nonlinear forecasting as a way of distinguishing chaos from measurement error in time series. *Nature*, **344**, April 19, 734-741.

[2] Sauer, T. (1992). Time series prediction by using delay coordinate embedding, *Non Linear Modeling and Forecasting*, edited by Casdagli and Ewbank, Sante Fe Institute, 175-195.

[3] Pineda, F. & Sommerer, J. (1992). Estimating generalized dimensions and choosing time delays: A fast algorithm, *Non Linear Modeling and Forecasting*, edited by Casdagli and Ewbank, Sante Fe Institute, 367-387.

[4] Kantz, H. (1992). Noise reduction by local reconstruction of the dynamics, *Non Linear Modeling and Forecasting*, edited by Casdagli and Ewbank, Sante Fe Institute, 475-491.

[5] Fraser, A. & Dimitriadis, A. (1992). Forecasting probability densities by using hidden Markov models, *Non Linear Modeling and Forecasting*, edited by Casdagli and Ewbank, Sante Fe Institute, 265-283.

[6] Gershenfeld, N. & Weigend, A. (1992). The future of time series: Learning and understanding, *Non Linear Modeling and Forecasting*, edited by Casdagli and Ewbank, Sante Fe Institute, 1-71.

[7] Weigend, D. & Rumelhart, G. (1992). Predicting sunspots and exchange rates with connectionist networks. *Non Linear Modeling and Forecasting*, edited by Casdagli and Ewbank, Sante Fe Institute, 395-432.[1]

[8] Grassberger, P. and Proccacia, I. (1983). *Physical review of letters.* **50**, 346-369

---

[1] This article felt that the simplex projection method relied too heavily on the embedding dimension. Their approach to prediction is one that is not as sensitive to the embedding dimension. Theirs has "an advantage over other prediction methods such as the simplex algorithm employed by Sugihara and May," (page 416).

CHAPTER 14

DEVELOPMENT OF MOBILE ROBOT WALL-FOLLOWING ALGORITHMS USING GENETIC PROGRAMMING

Robert A. Dain
Sierra On-Line
bob.dain@sierra.com.

ABSTRACT

This chapter demonstrates the use of genetic programming (GP) for the development of mobile robot wall-following navigation algorithms. Algorithms are developed for a simulated mobile robot that uses an array of range finders for navigation. Navigation algorithms are tested in a variety of different shaped environments to encourage the development of robust solutions, and reduce the possibility of solutions based on memorization of a fixed set of movements or solutions that capitalize on insignificant coincidental properties of the simulation environment. Results are presented and GP is shown to be capable of producing robust wall-following navigation algorithms that perform well in each of the test environments used.

INTRODUCTION

This work focuses on using genetic programming (GP) to develop mobile robot navigation strategies that will perform well in the real world. A simulated robot is used for learning, but the simulation is designed to closely approximate a real robot. The algorithms developed with GP can then be tested in the real world, on a real robot. Wall-following was selected for the initial problem domain because it is a fairly simple problem to set up and evaluate, and because it lays the groundwork for more complex problem domains, such as maze traversal, mapping, and full coverage navigation (i.e., vacuuming and lawn mowing). The development process for these behaviors is described, and the results of the experiments are presented.[1]

Genetic programming is attractive for this work for several reasons. First, it is a machine learning technology, so control strategies are emergent rather than predetermined. The GP environment is set up for a particular problem domain and it is left to "the unyielding weight of evolution," as Dr. John Koza puts it [1], to drive out solutions. Also, GP is a black-box system in that a-priori knowledge of the system under analysis is not required. Some method of ana-

---

[1] All experiments are conducted using a genetic programming application developed by the author.

where the original node was removed. Figure 14.3 shows an example of mutation.

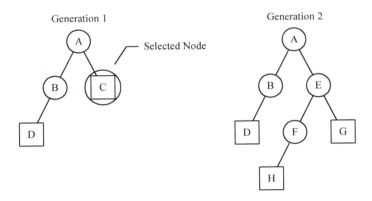

Figure 14.3 - The mutation operation.

Crossover involves selecting two individuals from the previous generation and selecting a node at random in each of them. The selected nodes, along with any sub-trees that exist below them, are exchanged between the two individuals, as shown below in Figure 14.4.

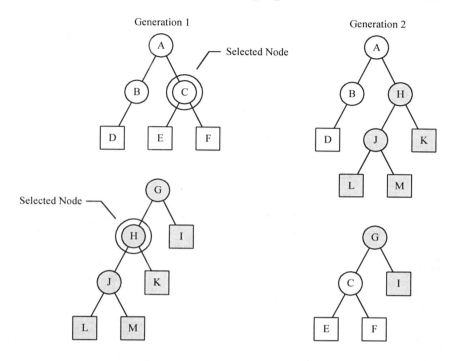

Figure 14.4 - The crossover operation.

As successive populations are created by applying these genetic operators, the average performance of the population tends to improve, and the performance of the best individual in the population can improve quite dramatically. There is no guarantee that GP will find an optimal solution (in less than infinite time), but a well thought out set of functions and terminals with a reasonable fitness test will usually produce good results.

PROBLEM DESCRIPTION

The goal of a wall-following robot is to navigate through open spaces until it encounters a wall, and then navigate along the wall at some fairly constant distance. The successful wall-follower should traverse the entire circumference of its environment without either staying too close, or straying too far from the walls.

A two-dimensional simulated environment is used to test the navigation algorithms developed by GP. The environment consists of four completely enclosed rooms with wall contours of varying complexity. Each individual algorithm from the GP population is tested in each of the four rooms, and scored on how well it performs the wall-following behavior. By testing each individual in a variety of different environments, GP is encouraged to develop general-purpose solutions rather than simply memorize a set of movements, or capitalize on insignificant coincidental properties of the simulator [2].

A screen display of the simulation environment is shown in Figure 14.5. The dark shaded areas represent the walls of the environment. The lightly shaded areas represent the desired path of the robot. The desired path is for display only – the robot has no visibility of these areas. All sensory input to the robot comes through the use of simulated range finders. The robot has eight range finders positioned at 45° increments about its center. The solid square in the center of each room designates the starting point of the robot in that room.

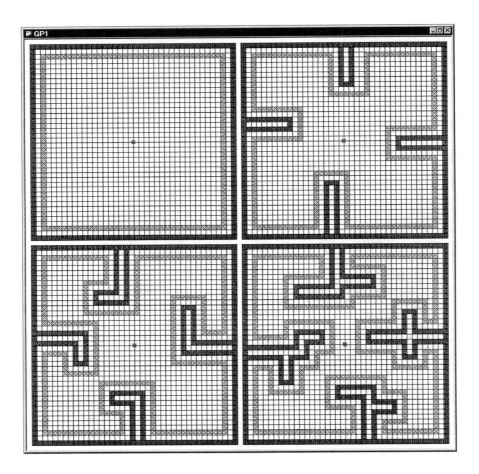

Figure 14.5 - The four room configurations used to develop the wall-following behavior.

Each room in the simulated environment consists of a 40 by 40 array of grid locations which are each 10 pixels square. The pixel size is used as the basic unit of movement.[2] The navigation algorithms are scored on how many of the lightly shaded "corridor" grid locations they can pass through – the more the better. The robot is allowed to wander between 10 and 20 units from the wall (the width of a corridor grid) without any performance demerits. GP calculates fitness based on the number of corridor grids which are not visited during the course of a run, and attempts to minimize the following fitness equation:

> 1000 * NumberOfCorridorGridsNotVisited +
> 100 * DistanceOfEndingPointFromCenterOfCorridor +
> TotalDistanceTravelled

NumberOfCorridorGridsNotVisited is the most significant aspect of fitness. A weighting factor of 1000 ensures that GP will remain motivated to navigate through corridor grids, even when there are only a few unvisited grid points remaining.

DistanceOfEndingPointFromCenterOfCorridor is the distance from the final resting point of the robot to the center of the nearest corridor grid (i.e., 15 pixels from the nearest wall). This term is used to kick-start the GP process. Without it, a robot that does not move at all (travels zero distance) will score higher than a robot that moves towards a wall but does not quite reach the corridor grids. In other words, this term encourages initial exploration by rewarding algorithms that move towards walls, even if they do not reach those walls. This provides the necessary scoring gradient in the early generations to encourage GP to develop individuals that leave their starting location and move towards the walls. A weighting factor of 100 is applied to this term so that it will eclipse TotalDistanceTravelled any time the robot does not reach the corridor grids.[3]

TotalDistanceTravelled is the least significant term in the fitness equation. It is included to discourage the navigation algorithms from traversing the corridor grids by spiraling haphazardly through the open spaces of the environment. A robot that navigates directly through the center of the corridor grids will score higher than one that meanders back and forth across the grids.

---

[2] Robot position is displayed at the nearest pixel, but is not constrained to pixel centers.

[3] This term can also eclipse (or at least compete with) TotalDistanceTravelled if a robot moves away from the walls after passing through one or more corridor grids. While that is not the intent of this term, the effect does not seem to hinder performance and may even be beneficial.

GENETIC PROGRAMMING ENVIRONMENT

As described above, genetic programming solves problems by combining and organizing a predefined set of functions and terminals. This set of functions and terminals provide the fundamental domain knowledge which allows GP to discover solutions. The functions and terminals defined for the wall-following experiments presented herein are defined below.

Do2 (Arga, Argb) This function provides for sequential operations. It evaluates argument Arga, then it evaluates argument Argb. The return value is the result of argument Argb. Multiple sequential steps can be accomplished by nested Do2's.

WhileInCoridorRange (Arga) This function checks each of the sensor distance values, selects the minimum value (i.e., the closest wall) and determines if it is within the allowable wall-following range. If the robot is out of range of a wall, this function returns the value 0.0. Otherwise, it evaluates argument Arga continuously (as long as the robot is within acceptable range of the nearest wall) and returns the value obtained from the last evaluation of argument Arga. The range of values that satisfies the "InCorridorRange" condition is defined as 10 to 20 units, inclusive.

WhileTooCloseToWall (Arga) This function is identical to WhileInCoridorRange except that it evaluates the argument as long as the robot is too close to a wall. The range of values that satisfy this function is less than 12 units. 12 is used rather than 10 to provide a hysteresis band between the in and out of range conditions.

WhileTooFarFromWall (Arga) This function is also identical to WhileInCoridorRange except that it evaluates the argument as long as the robot is too far away from the closest wall. The range of values that satisfy this function is greater than 18 units.

IfConvexCorner (Arga, Argb) This function tests for the condition of the range sensor at angle 90° reading too far to the nearest wall, and the range sensor at angle 135° reading an acceptable distance. This condition occurs when the robot is following a wall that drops away in a convex corner. If this condition is satisfied, the function evaluates argument Arga and returns its value. Otherwise, it evaluates argument Argb and returns its value.

MoveForward() This terminal causes the robot to move forward five units. The return value is 1.0 if the robot completes its move without colliding with a wall, and 0.0 if a collision occurs. If the robot collides with a wall, movement stops at that point (i.e., the robot is not allowed to move into or through the walls). These return values are arbitrary and were selected to coincide with the C/C++ integral values used to represent Boolean true and false.

TurnRight(), TurnLeft() These terminals cause the robot to rotate 45° to the right or left, respectively. The return value from these terminals is the angle that the robot ends up facing after the rotation.

TurnTowardsClosestWall() This terminal checks each of the current range sensor values and selects the lowest value (i.e., the closest wall). It then rotates the robot such that the robot is pointed in the direction of the sensor that had the lowest reading. Note that the robot turns towards the perceived closest wall, and not necessarily towards the actual closest wall. The return value is the new direction of the robot.

TurnAwayFromClosestWall() This terminal is similar to TurnTowardsClosestWall, except that the robot rotates to face the opposite direction of the perceived closest wall.

TurnParallelToClosestWall() This terminal is also similar to TurnTowardsClosestWall, except that the robot rotates such that its 90° sensor is pointed in the direction of the perceived closest wall.

Population size was set to 500 individuals, and experiments typically ran for 200 to 300 generations. The probability values for reproduction, crossover, and mutation were set to 39%, 60% and 1%, respectively, for all experiments. All experiments invoked an option to force the best performing individual to reproduce at least once into the next generation. All experiments were performed on Pentium-100 PC's.

RESULTS

The genetic programming engine was, indeed, able to organize the functions and terminals described above into fairly proficient wall-following algorithms. This section describes a typical evolutionary cycle. The initial generations of each run, as expected, performed quite poorly. The robot paths in Figure 14.6 (shown as dotted lines) are typical of early generations. The robot manages to maneuver into the wall-following corridor and claim a few grid points, but has not yet figured out how to turn and follow the corridor. (The target corridor shading has been removed for clarity.)

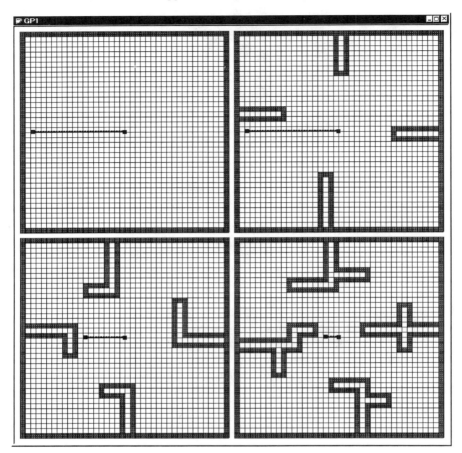

Figure 14.6 - Typical result of early generations in a GP run.

The GP-generated code that produced the behavior displayed in Figure 14.6 is shown below. The line numbers are for reference only, and the indentation represents the calling hierarchy. This particular solution capitalizes on the hysteresis band in the WhileTooFarFromWall function to skim the edge of a wall following corridor in the upper right room.

It begins with a WhileTooFarFromWall loop, with a Do2 function as the body. The robot starts in a too-far-from-wall position, so the Do2 is evaluated. The first argument of the Do2 is a TurnAwayFromClosestWall terminal which sets the direction of the robot to the left in all cases. The second argument of the Do2 is another WhileTooFarFromWall loop with a MoveForward terminal as its body. This loop repeats until the too-far-from-wall test fails (i.e., the robot ends up within allowable wall-following range). The inner WhileTooFarFromWall loop terminates at that point. The Do2 is complete, so the outer WhileTooFar-FromWall loop evaluates the position of the robot. Finding that the robot is not too-far-from-wall, this loop terminates and the program ends with a total of 13 hits out of the possible 933.

```
0   WhileTooFarFromWall
1       Do2
2           TurnAwayFromClosestWall
3           WhileTooFarFromWall
4               MoveForward
END
```

GP soon begins to develop solutions that exhibit rudimentary wall-following behavior, as shown in Figure 14.7. These solutions typically perform reasonably well in the square room, and claim a few corridor grids in the more complex rooms by ricocheting around the environment.

The code for the individual displayed in Figure 14.7 is shown below. It begins with a WhileTooFarFromWall loop with a Do2 as its body. The arguments of the Do2 are a MoveForward and a Do2 with two WhileInCoridorRange loops as arguments. While the robot is out of corridor range the WhileInCoridor-Range tests will fail. The result is that the robot will move forward until it finds itself within corridor range of a wall. When the MoveForward on line 2 moves the robot into a wall-following corridor, the WhileInCoridorRange functions are able to execute. The first (on line 4) turns the robot away from the closest wall repeatedly until the maximum loop counter is exceeded.[4] The second (on line 6) employs a Do2 to turn the robot parallel to the closest wall and move forward. These two steps are repeated until the robot moves out of range, or the loop times out.[5] When either of those conditions occurs, another iteration of While-TooFarFromWall (on line 0) is made. If the robot is out of range and the loop counter is not maximized, the process repeats. This algorithm passes through a total of 275 corridor grids.

---

[4] Actually these loops terminate immediately if the robot does not move during the course of evaluating the function body. This feature was added to improve the performance of the simulator.

[5] The instruction pair, TurnParallelToClosestWall and MoveForward, are capable of navigating a concave turn because the robot gets closer to the wall it is approaching than the wall it is following before it goes out of corridor range. The instruction pair is not capable, however, of negotiating a convex corner – the robot simply moves out of range. This is the reason the robot follows the walls briefly and then loses them.

```
0    WhileTooFarFromWall
1        Do2
2            MoveForward
3            Do2
4                WhileInCoridorRange
5                    TurnAwayFromClosestWall
6                WhileInCoridorRange
7                    Do2
8                        TurnParallelToClosestWall
9                        MoveForward
END
```

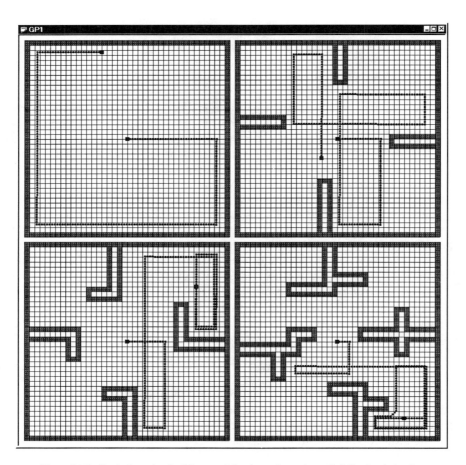

Figure 14.7 - Typical result of a GP run as it begins to learn the wall-following behavior.

In almost every run GP is able to generate algorithms that perform reasonably well. The individual shown in Figure 14.8 makes a couple of errors, but for the most part it is exhibiting the desired behavior.

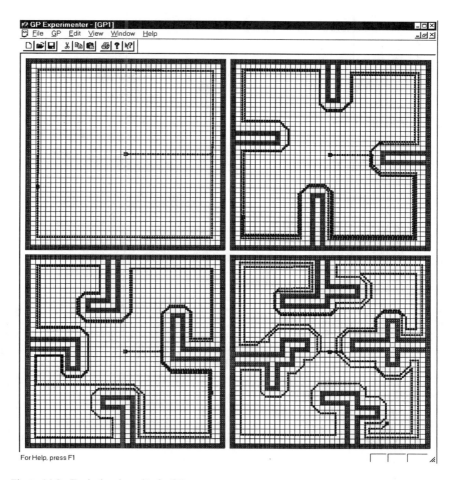

Figure 14.8 - Typical end result of a GP run.

The code for the individual displayed in Figure 14.8 is shown below. Analysis of this code (which is typical of the complexity of GP-generated algorithms) is left as an exercise for the reader.

```
0   WhileTooFarFromWall
1   WhileTooFarFromWall
2   Do2
3   Do2
4   WhileTooCloseToWall
5   WhileInCoridorRange
6   TurnParallelToClosestWall
7   Do2
8   WhileInCoridorRange
9   Do2
10  WhileTooCloseToWall
11  WhileTooFarFromWall
12  MoveForward
13  Do2
14  Do2
15  WhileTooCloseToWall
16  TurnParallelToClosestWall
17  Do2
18  Do2
19  WhileTooCloseToWall
20  WhileInCoridorRange
21  TurnParallelToClosestWall
22  Do2
23  WhileInCoridorRange
24  Do2
25  WhileTooCloseToWall
26  WhileTooFarFromWall
27  MoveForward
28  Do2
29  Do2
30  WhileTooCloseToWall
31  TurnParallelToClosestWall
32  Do2
33   MoveForward
34   nTowardsClosestWall   ·
35   Do2
36   MoveForward
37   WhileInCoridorRange
38   Do2
39  TurnTowardsClosestWall
40  Do2
41  TurnParallelToClosestWall
42  MoveForward
43  MoveForward
44  WhileInCoridorRange
45  TurnAwayFromClosestWall
```

46 Do2
47 MoveForward
48 WhileInCoridorRange
49 Do2
50 TurnTowardsClosestWall
51 Do2
52 TurnParallelToClosestWall
53 MoveForward
54 MoveForward
55 WhileInCoridorRange
56 TurnAwayFromClosestWall
END

CONCLUSION

Genetic programming has demonstrated the capability of generating wall-following navigation algorithms from the functions and terminals provided. More generally, the experiments conducted over the course of this project have shown the feasibility of using genetic programming to develop mobile robot navigation algorithms. This project lays the foundation for planned follow-on projects of maze traversal, map generation, and full-coverage area traversal. The results of this project show enough promise to warrant further research into these more complex tasks.

REFERENCES

[1] Koza, J. R. (1992). *Genetic programming: On the programming of computers by means of natural selection.* Cambridge: MIT Press.

[2] Reynolds, Craig W. (1994) *Evolution of obstacle avoidance behavior: Using noise to promote robust solutions.* Advances in Genetic Programming, pages 221-241. Cambridge: MIT Press.

CHAPTER 15

HYDROCYCLONE MODEL USING GENETIC PROGRAMMING

Barry Weck and Charles L. Karr
Department of Aerospace Engineering and Mechanics
University of Alabama
Box 870280
Tuscaloosa, AL 35487-0280
e-mail: bweck@eng.ua.edu and ckarr@coe.eng.ua.edu

ABSTRACT

Developing computer models from system data is an important problem. In general, this is a problem of system identification which is the problem of determining the mathematical equations that describe the fundamental behavior of the system being addressed. This problem arises in a wide variety of engineering disciplines, and is important both in process simulation and process control. This chapter describes the use of genetic programming for solving a particular system identification problem from the minerals industry. Specifically, genetic programming is used to determine a functional relationship that describes the behavior of a hydrocyclone, a device used in the minerals industry to achieve mineral separations.

INTRODUCTION

The problem of system identification is the problem of determining the mathematical equations that describe the fundamental behavior of the system being addressed. System identification problems must be solved to achieve a wide variety of goals in an even wider variety of fields. For instance, system identification problems are prominent in the areas of (1) simulation, (2) scheduling and resource allocation, and (3) process control. Such problems exist in almost every engineering and scientific discipline, as well as in business and industrial applications.

The techniques and approaches to solving system identification problems are almost as varied as the disciplines in which these problems appear. These methods range from the traditional first-principles approach to developing artificial intelligence techniques such as neural networks and fuzzy logic. In a first principles approach, the fundamental physics of a system are modeled using governing equations. In the artificial intelligence-based techniques, data depicting the behavior of the system are presented to the technique of choice, and a mathematical relationship describing the characteristics of the system are developed. Certainly, there are other approaches, but since there are almost as many approaches as there are system identification problems, this chapter will focus on a specific area. Namely, the focus of this chapter is on system identification in the minerals industry. The example used is that of developing a computer model of a hydrocyclone separation device.

Here, the system identification problem will be posed as follows. Determine the behavior of a system, $f(x)$, of a particular system given only data depicting system input, x, and the subsequence system output, f. Thus, for the hydrocyclone problem, the input and operating parameters will be considered known, as well as the resulting split size (see discussion on hydrocyclone later in this chapter). Based on a limited number of such data values, the system identification problem is to determine a generalized relationship that can be used to predict the split size based on the input and operating parameters.

Traditional first-principle models are difficult to develop for systems in the minerals industry because the complex chemistry and physics associated with these processes are not fully understood [1]. This is particularly true for hydrocyclones which are characterized by three-phase, three-dimensional flows. However, there is little doubt that the minerals industry can benefit from the application of computer models through the successful solution of system identification problems in the areas of equipment design and process control. Unit processes, including froth flotation, column flotation, grinding, and leaching are currently operated at less than optimum conditions, and the development of accurate computer models represents a major step toward rectifying this situation as discussed in a paper by Karr, Yeager, and Stanley [2]. Efficient (if not optimized) computer models can be used to achieve improved efficiency via their incorporation into adaptive control systems, automated equipment design systems, and scheduling algorithms. Because there is tremendous potential for economic improvements in the separation industry, researchers have begun to focus on the development of computer models of mineral processing systems which are based on artificial intelligence techniques [3, 4]. Two approaches have been found to be particularly effective in developing data-driven computer models: (1) neural networks and (2) fuzzy mathematics with genetic algorithms. Unfortunately, both of these methods have shortcomings.

Neural networks are promising tools for model development in the separation industry. In fact, at first consideration, they seem like a panacea for modeling because the developer is basically required to understand nothing about the relationships between the system parameters. Unlike traditional modeling techniques, neural networks are not "programmed," rather they are "trained" [5]. Neural networks are presented with input-output data collected from the system and use this data to become proficient predictors of the future input/output response of the system. Although this approach is inviting for a variety of reasons, there is a drawback. Once a neural network has been trained, it is truly a "black box;" the neural network model provides no insight into its predictions. In the complex systems found in the separation industry where safety is a key issue, this limitation can be inadequate. Additionally, this limitation can deem neural networks ineffective in traditional process control algorithms.

Fuzzy mathematics offers a nice alternative to neural network models. Fuzzy mathematics provides a mechanism by which the "rule-of-thumb" approach often used by humans to solve difficult problems can be incorporated into a set of traditional production rules [6]. The resulting production rules are linguistic in nature, and thus, can be easily comprehended and interpreted by a human. The linguistic rules and membership functions can then potentially be used to understand the mechanics of the system being modeled. The development of a fuzzy linguistic model is complicated only by the difficulty associated with selecting efficient rules and membership functions. There are however, automated approaches for tuning fuzzy systems.

The rules and membership functions required by a fuzzy linguistic model can be discovered using a genetic algorithm. Genetic algorithms are search algorithms based on natural genetics, and these innovative techniques have been used successfully to solve a wide variety of search problems [7]. Genetic algorithms possess a number of qualities that make them inviting for use in a self-generating fuzzy linguistic modeling system [8]. Despite the relative success of using genetic algorithms to evolve the rules and membership functions required in fuzzy systems, effectively tuning a fuzzy system can be a daunting and time-consuming task.

In this chapter, an alternative approach is considered. Here, genetic programming is used in an attempt to solve a system identification problem related to a hydrocyclone separation device. Genetic programming [9] is a genetic algorithm-based approach to machine learning in which entire programs are generated based exclusively on data. In this way, they are similar to neural networks. However, there is an important fundamental difference in the results of neural networks and genetic programming. Genetic programming results in a definite equation containing function calls described by the user. The previous chapter provides an overview of genetic programming, thus such a description is not repeated here. From the previous chapter, it is clear that complex functions developed by the user and specific to the application at hand can be incorporated into the genetic programming solution. This capability can be quite important and can help prevent the "re-invention of the wheel," so to speak.

In the current chapter, the hydrocyclone system identification problem is described. Next, the key characteristics of the genetic programming search are described. Then, results are presented demonstrating the effectiveness of genetic programming in solving the particular system identification problem. Finally, conclusions are presented.

HYDROCYCLONE DATA FOR FUZZY MODEL DEVELOPMENT

Hydrocyclones are commonly used for separating slurries in the mineral processing industry [10], and are used extensively to perform separations in other industries as well. Hydrocyclones utilize centrifugal forces to accelerate the settling rate of particles. Since their mechanical structure is simple, durable, and relatively inexpensive, hydrocyclones are one of the most popular mineral separating devices found in industry. They are used in closed-circuit grinding, de-sliming circuits, de-gritting procedures, and thickening operations. Figure 15.1 shows a schematic of a typical hydrocyclone. The lower part of the hydrocyclone has a conical shape with an opening at the apex, allowing the coarse or heavier particles to be removed via the underflow. The top section of the hydrocyclone is cylindrical and is closed with the exception of an overflow pipe called a vortex finder. This vortex finder prevents the mineral sample from going directly into the overflow, while allowing the fine particles to remain in the hydrocyclone. The actual mineral separation occurs in the cylindrical section due to the existence of a complex velocity distribution that carries the fine particles out the top and the coarse particles to the apex.

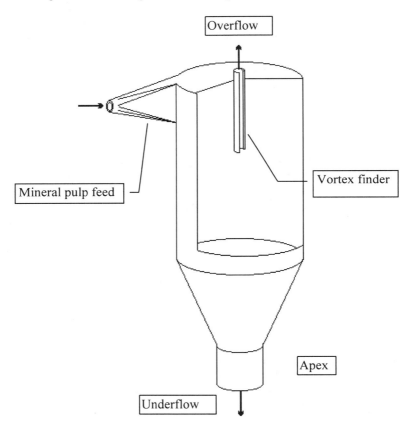

Figure 15.1 - A typical hydrocyclone has a conical lower section and a cylindrical upper portion.

There have been numerous efforts to model hydrocyclones. Plitt [11] developed a statistical model to predict the split size of the hydrocyclone, d_{50}. The split size is that size particle that has a 50 percent chance of exiting in either the overflow or the underflow. Plitt's model has proven to be quite robust and is often cited in the literature. Other fairly effective models have been developed by various researchers [12,13,14]. Despite the numerous efforts to model hydrocyclones, none of the models have been universally adopted. Most of the models are either too computationally intensive or are applicable to a limited range of hydrocyclone designs. Therefore, the intent here is to develop a robust, fuzzy-based hydrocyclone model. To facilitate this hydrocyclone modeling effort, hydrocyclone performance data has been acquired from the U.S. Bureau of Mines. This data pertains to a wide range of hydrocyclones used on a variety of mineral samples including iron, phosphate, and copper. A computer model of a hydrocyclone must consider several input parameters to accurately compute a value of d_{50}. Figure 15.2 is a schematic of a hydrocyclone model where:

D_c = diameter of the hydrocyclone,

D_i = diameter of the slurry input,

D_o = diameter of the overflow,

D_u = diameter of the underflow,

h   = height of the hydrocyclone,

Q   = volumetric flow rate into the hydrocyclone,

ϕ = percent solids in the slurry input,

ρ = density of the solids,

d_{50} = hydrocyclone split size.

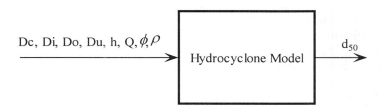

Figure 15.2 – A typical hydrocyclone computer model receives various input parameters in an attempt to predict the d_{50} size.

The data available for the model development consists of input-output pairs. The inputs consist of the variables shown entering the model in Figure 15.2. The output is the associated d_{50}. The data is segregated into two distinct groups: (1) a training set, and (2) a test set. In the training data set, the desired output data are provided to the fuzzy system so as to update the fuzzy system parameters in attempts to develop an accurate model. The desired output is not presented to the fuzzy model in the test set, so this test set serves as a test suite for measuring the performance of a trained system.

GENETIC PROGRAMMING PARTICULARS

In his 1992 book [9], John Koza recommends six steps that should be followed to successfully complete a genetic programming implementation. In this section, the particulars of these six steps as they apply to the hydrocyclone system identification problem will be discussed.

1. Identify the Terminals

In genetic programming applications, there are two key elements: terminals and functions. Terminals are the things that can appear at the terminal points in the graphical (or tree) representations of possible computer programs (see Chapter 14 for a detailed discussion). Examples of terminals include variables and constants.

Recognize that, theoretically, the solution space in a genetic programming problem is all possible computer programs. Contemplating this point makes the task appear quite daunting, if not impossible. Thus, the size of the search space is limited some by reducing the number of possibilities. The first step in doing so is to limit the terminals that are considered in the candidate computer programs. For instance, in the current problem, it is unlikely that the computer programs that successfully simulate the performance of a hydrocyclone will require the use of alpha characters. Thus, these can be eliminated from the pool of possible terminals, thereby reducing the size of the search space.

The terminals selected in the current problem are all floating point values (which may serve as constants in any equations) and all of the input and output variables in the hydrocyclone problem:

1. D_c, the diameter of the hydrocyclone,

2. D_i, the diameter of the slurry input,

3. D_o, the diameter of the overflow,

4. D_u, the diameter of the underflow,

5. h, the height of the hydrocyclone,

6. Q, the volumetric flow rate into the hydrocyclone,

7. ϕ, the percent solids in the slurry input,

8. ρ, the density of the solids, and

9. d_{50}, the hydrocyclone split size.

2. Identify a Set of Functions

This can represent a tricky issue. Since there have been a number of empirical models developed for simulating hydrocyclone performance, the temptation here was to use the functions employed in some of the more accurate models. However, for the genetic programming approach to work in general on system identification problems, this cannot be considered a viable approach. The approach adopted here was to view the modeling equation as being a function that could perhaps be represented as a Fourier Series. Therefore, the functions selected for the current problem were:

1.   addition,
2.   subtraction,
3.   multiplication,
4.   division (with the special requirement that division by zero results in zero),
5.   sine,
6.   cosine, and
7.   exponentiation.

3. Sufficiency and Closure

Actually, this step is directly related to step 2 above of selecting the functions to be considered. One of the problems associated with limiting the pool of functions from which genetic programming can assemble a computer program is that this pool may be too restrictive; the necessary functions might not be included in the candidate pool. Thus, sufficiency involves ensuring that the set of terminals and the set of primitive functions are capable of expressing a solution to the problem at hand. Unfortunately, ensuring the extent to which the functions and terminals are sufficient to solve the problem is often difficult to judge until some genetic programming runs have been made.

The idea of closure is related to the problem of generating computer programs that are feasible, or that will compile. Most programmers (even highly skilled ones) generate computer programs that contain fundamental syntax errors. Certainly, this phenomenon can occur when a genetic algorithm is "randomly" generating codes by mixing and matching pieces of existing codes. Thus, in genetic programming, closure must be ensured. Closure dictated that each of the functions in the function set must be able to accept, as its arguments, any value and data type that may possibly be returned by any function in the function set. In this way, genetic programming will not generate infeasible codes; ineffective perhaps, but not infeasible.

4. Identify a Measure of Quality

The search conducted in genetic programming is performed by a genetic algorithm. Recall that genetic algorithms use as their only driving force, a fitness function which defines just how effective (or ineffective) a candidate solution is in solving the problem at hand. In the hydrocylcone identification problem, the effectiveness of a candidate solution is directly related to how well it predicts the split size of a particular set of hydrocyclones. Given this, the fitness function (or measure of quality) is:

$$f = \sum_{i=1}^{i=N} \left[(d_{50}^{\ data})_i - (d_{50}^{\ calculated})_i \right]^2$$

where N is the number of data points available, d_{50}^{data} is the split size appearing in the data set, and $d_{50}^{calculated}$ is the split size computed for a particular set of input values. Genetic programming is used to compute candidate solutions (computer programs depicting equations for calculating the split size). Each candidate solution is then used to compute a d^{50} size for N=200 input vectors (values of D_c, D_i, D_o, D_u, h, Q, ϕ, and ρ). Naturally, the more accurate the model, the closer the value of f is to 0. Thus, the genetic algorithm in the genetic programming code has a value that it can strive to minimize. If it finds a candidate solution that has $f = 0$, then that solution predicts d^{50} values that exactly match all 200 data values.

5. Select Values to Control the Run

As with a standard genetic algorithm application, applying genetic programming to a problem requires the determination of some variable values. Along with those parameters typically associated with a genetic algorithm such as population size and probability of crossover, there are some parameters that are specific to genetic programming.

The pertinent variables and the values used in the current problem are:
1. Population size = 500,
2. Maximum number of generations = 25,000,
3. Initial depth of the randomly generated initial structures = 5-10,
4. Maximum depth of any structure allowed in the genetic programming run = 20, and
5. Number of data points used in the evaluation of fitness = 200.

6. Select a Stop Criterion

In static problems such as the one considered here, selecting a stop criterion is a rather straightforward task. Here, the genetic programming is continued for a maximum number of generations (25,000), or until a solution is found that exactly models the data presented it ($f =0$). But, in other problems, appropriate stopping criteria are not so readily apparent. For instance, in real-time control systems, there are situations wherein the time to search is limited, the search must be terminated when something changes in the plant, or a variety of other equally acceptable criteria.

Conclusions

Following the six steps above proves to be quite useful in implementing genetic programming for solving a particular problem. Although each of these steps results in facets of the run that can dramatically affect the results obtained, the most important ones appear to be the selection of the terminals and functions, and their relationship to the ideas of sufficiency and closure.

RESULTS

A genetic programming simulation was run according to the guidelines and parameters set forth in the preceding sections of this chapter. Since the objective was to consider the effectiveness and robustness of genetic programming as a system identification tool, the hydrocyclone problem was considered in stages.

The split size (d_{50}) of a hydrocyclone is generally considered to be a function of eight variables: $d_{50}= f(D_c, D_i, D_o, D_u, h, Q, \phi, \rho)$. However, a system identification problem with eight variables is terribly challenging. Additionally, many mineral separation plants are not interested in the functional relationship between all eight of the independent parameters because they have no control over some of them. Furthermore, some of these independent variables are constant in a particular situation. Thus, in the initial phase of this work, the split size was considered to be a function of but three variables: $d_{50}= f(D_c, D_i, D_o)$. Figures 15.3 and 15.4 provide an indication of the effectiveness of using genetic programming for this problem.

Figure 15.3 is a 45° plot in which the model predicted value of d_{50} is plotted against the actual data value. If the model exactly matched the data, all of the points would fall on a 45° line. It can be seen from the figure that for the majority of the split size values, the model does an adequate job of predicting. However, for the larger split sizes, the model performance degrades. Interestingly enough, this phenomenon is observed in many hydrocyclone models.

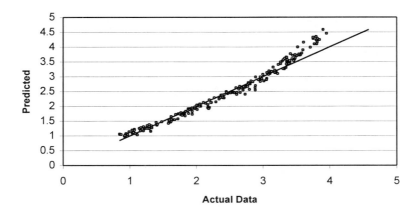

Figure 15.3 – The 45° plot above shows that for most split sizes, the model discovered using genetic programming was quite effective.

Figure 15.4 shows basically the same information as is depicted in Figure 15.3, only in a different form. Figure 15.4 is a bar graph indicating the number of model predicted values that were accurate to a given accuracy. Here, it can be seen that most of the values are reasonably accurate. Be advised that some of the large percent errors are due to the fact that the d_{50} values are quite small.

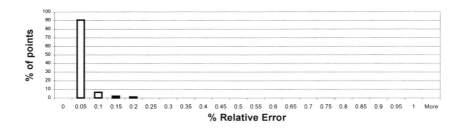

Figure 15.4 – The above bar graph demonstrates the fact that the majority of the d_{50} values predicted by the model are easily within the accepted engineering accuracy of 10%.

Phase two of this research effort considered additional independent variables. At this point, the hydrocyclone split size was considered to be a function of four variables: $d_{50} = f(D_c, D_i, D_o, \phi)$. Figures 15.5 and 15.6 provide an indication of the effectiveness of using genetic programming for this problem. Figure 15.5 is a 45° plot while Figure 15.6 is the corresponding bar chart. Notice that although the hydrocyclone model discovered using genetic programming is not as accurate in the previous example, it still performs acceptably. A further increase in parameters would have associated with it a further degradation in performance. Efforts are ongoing to improve the performance of genetic programming in this problem domain.

Figure 15.5 – The 45° plot above shows that the model discovered using genetic programming was effective.

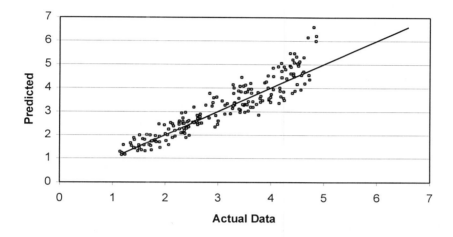

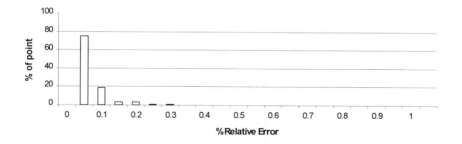

Figure 15.6 – The above bar graph demonstrates the fact that although most of the d_{50} values predicted by the model are within the accepted engineering accuracy of 10%, many are not. This is due in part to the fact that so many of the d_{50} sizes are quite small, and therefore, small differences correspond to large percent errors.

CONCLUSIONS

In this chapter, genetic programming was used to solve a rather difficult system identification problem from the separation industry. A modeling equation was discovered that allowed for the accurate prediction of split size as a function of up to four variables. Efforts are ongoing to extend the work so that the relationship between the split size and all eight variables generally considered important in this problem can be determined.

The results presented here are quite promising. System is a very difficult yet very important problem in a wide range of industries. If genetic programming can effectively improve or simplify the process of developing modeling equations, then the economic impact can be substantial.

REFERENCES

[1] Kelly, E. G. & Spottiswood, D. J. (1982). *Introduction to mineral processing.* New York, NY: John Wiley & Sons.

[2] Karr, C. L., Yeager, D., & Stanley, D. A. (1994). Tuning empirical computer models using genetic algorithms. *Proceedings of the Society for Computer Simulation's Summer Meeting*, 236-241.

[3] Karr, C. L. (1993a). Strategy for adaptive process control for a column flotation unit. *Proceedings of the Fifth Workshop on Neural Networks*, **SPIE 2204**, 95-100.

[4] Karr, C. L. (1993b). Optimization of a computer model of a grinding process using genetic algorithms. In H. El-Shall, B. Moudgil, & R. Weigel (Eds.), *Beneficiation of Phosphate: Theory and Practice* (pp. 339-345). Littleton, CO: Society for Mining, Metallurgy, and Exploration, Inc.

[5] Wasserman, P. D. (1989). *Neural computing.* New York, NY: Van Nostrand Reinhold.

[6] Zadeh, L. A. (1975). Outline of a new approach to the analysis of complex systems and decision processes. *IEEE Transactions on Systems, Man, and Cybernetics*, **SMC-3**, (1), 28-44.

[7] Goldberg, D. E. (1989). *Genetic Algorithms in Search, Optimization, and Machine Learning.* Reading, MA: Addison-Wesley Publishing Company, Inc.

[8] Karr, C. L. (1991). Genetic algorithms for fuzzy controllers. *AI Expert*, **6** (2), 26-33.

[9] Koza, J. R. (1992). *Genetic programming: On the programming of computers by means of natural selection.* Cambridge: MIT Press.

[10] Willis, B.A. (1979). *Mineral processing technology.* Toronto: Programon Press.

[11] Plitt, L. R. (1976). A mathematical model of the hydrocyclone classifier. *CIM Bulletin*, **69**, 114-123.

[12] Bradley, D. (1965). *The hydrocyclone*. Elmsford, NY: Pergamon Press.

[13] Fahlstrom, P. H. (1963). Studies of the hydrocyclone as a classifier. *International Mineral Processing Congress*, **6**, 87-114.

[14] Lynch, A. J., & Rao, T. C. (1975). Modeling and scale-up of hydrocyclone classifiers. *International Mineral Processing Congress*, **11**, 245-269.

CHAPTER 16

WHAT CAN I DO WITH A LEARNING CLASSIFIER SYSTEM?

H. Brown Cribbs, III and Robert E. Smith
Department of Aerospace Engineering and Mechanics
The University of Alabama
Box 870280
Tuscaloosa, AL 35487-0280
e-mail: brown@galab3.mh.ua.edu

ABSTRACT

The learning classifier system (LCS) is an application of the genetic algorithm (GA) to machine learning. Artificial neural networks (ANNs) perform mappings of input vectors to outputs much the same way a LCS does. This chapter introduces the LCS paradigm and provides literature references for future investigation. Through the use of LCS principles, an ANN becomes a variable structure production system, capable of making complex input-output mappings that are similar to a LCS.

The evolutionary process of a single ANN facilitates a broad understanding of how evolution may help rule-based (or neuron-based) systems. An evolutionary approach to ANN structure is reviewed. Its similarities to the LCS are discussed.

A simple extension to Smith and Cribbs' (1994) and Cribbs' (1995) work in an ANN and LCS analogy is presented. The experiment presented removes the nonlinearity of the ANN's output layer to assess the nonlinear effects of the GA's partitioning within the hidden layer. The results indicate that GA-induced nonlinearity actively participates in the solution of a difficult Boolean problem – the *six multiplexor problem.*

INTRODUCTION

Learning classifier systems (LCSs) were developed by Holland in 1978 (Holland & Reitmann, 1978). Since its advent, the LCS has become a favorite of complex systems researchers and economists for its unique structure (Holland, 1993; Wilson, 1986). The use of genetic algorithms (GAs) allows the LCS to refine its rule-base, which improves the LCS's performance.

Artificial neural network (ANN) research began almost at the birth of electronic computers, but it was not until 1986 that the backpropagation training method emerged (Rumelhart, Hinton, & Williams, 1986). Since that time, the ANN's popularity has grown; subsequently, proofs of its abilities as a *universal approximator* of arbitrary precision have emerged in due course (Hornik et al., 1989; Cybenko, 1988).

Wilson (1990) presents an interesting extension to perceptron networks-- genetically evolved input partitions. Perceptrons (Rosenblatt, 1958) represent one of the simplest ANN types. Unfortunately, the work of Minsky and Papert (1969) showed that perceptrons were not capable of learning *linearly insepara- ble* tasks. Wilson's work (1990) provides a workable solution to linear insepa- rable problems and shows that input partitioning (input selection) adds another degree of nonlinearity to perceptron networks. Working on this premise, Smith and Cribbs (1994) used Wilson's perceptron experiment as a starting point to investigate analogies between multi-layer, feedforward ANNs and LCSs.

The basic LCS paradigm does much the same task as Wilson's 1990 work (Smith & Cribbs, 1994). Linearly inseparable classification tasks have become a favorite of LCS researchers and ANN researchers alike. This chapter intro- duces the basic principles of the LCS and mentions many advanced features to complete the reader's exposure to the LCS. After the introduction to the LCS, the discussion turns to the similarity in features of the LCS and ANNs.

LEARNING CLASSIFIER SYSTEMS

A Learning Classifier System (LCS) is a rule-based system that learns by in- teracting with its environment. The LCS observes its environment and notes regularities within that environment. From these observations, the LCS forms rules that dictate how the LCS acts.

LCS rules are linguistic in nature and can be thought of as IF-THEN state- ments. For instance, a rule to control a robot wandering about a room might be,

> **if** an object is to my right **AND** in front of me
> **then** turn left.

This sort of rule may be written in a shorthand notation. Ease of storage and manipulation within the computer motivate shorthand. A simple notation is to drop the **if** and **then** and simply list the conditions and action(s) separated by a slash. The rule above might appear in shorthand as,

$$\# \# \# 1 \ 1 / 1 \ 0 \ 0.$$

The left-hand side (LHS) of the rule above is called the *condition* under which the rule applies. The right-hand side (RHS) of the rule is the *action* the rule advocates. The condition side of the rule also has three fields with # (hash) characters in addition to the two 1's. The # is a special operator implementing the Boolean "don't care" operation. The two ones on the LHS are said to be defined bits, i.e., they denote the necessary conditions for the rule to be applicable. In the case of the rule above, the first one may be taken to mean, "an object is present to my right," and the second one similarly denotes the presence of an object in front of the robot. The RHS may be encoded by assuming there are three discrete actions: turn left, turn right, and move forward. The above rule would then have the action of "turn left, not right, and don't move."

The next few sections provide a brief introduction to LCSs, included are explanations of the various components, order of operation, and general issues surrounding their use. This introduction is intended for the LCS novice and for those unfamiliar with the LCS paradigm.

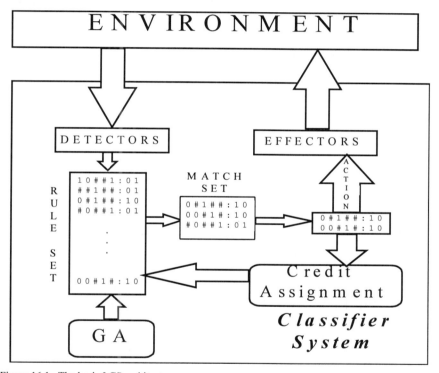

Figure 16.1. The basic LCS architecture.

How Does an LCS work?

An LCS is a *variable architecture learning system*, which uses a genetic algorithm (GA) to occasionally change its operating principles (rules) to improve its performance. The LCS obtains information about its environment from the *detectors*, and maps those inputs onto the rule conditions. The rules whose conditions match the detector's state are then considered for action.

Traditional LCSs use *discrete actions*, that is to say, actions are selected from the set of all available actions to the LCS. Actions are selected relative to a rule's *merit*. Rules use their merit to *bid* for the right to take action. The final action is selected based on merit, originally termed *strength*. Modern LCSs, like Wilson's XCS (1995), use metrics based on rule accuracy, rule confidence, or consensus (Wilson, 1994; Wilson, 1995; Smith & Cribbs, 1994).

The final action is determined via competition of all matching rules. The subset of rules advocating the same action with the highest combined merit is generally selected. Many action selection schemes have been developed, but most use a form of noisy auction. Noise is added to the bids (cumulative merit) to allow for exploration of the action space by stochastically selecting rules early on and slowly degrading the noise level. One method of doing this is to key the noise level to a Boltzmann distribution,

$$\Pr(a_i) = \frac{e^{-\frac{m_i}{T}}}{\displaystyle\sum_{\forall k \in A} e^{-\frac{m_k}{T}}} \qquad (16.1)$$

that decays as time progresses. Equation 16.1 represents the probability of action, a_i being selected based on its *merit, m_i*, in light of all the possible actions available for this time step.

The Predominant Components of an LCS

No matter what "flavor" of LCS, there are similarities. For instance, rules are necessary, and some form of rule discovery mechanism is required. In addition to these two items, facilities for credit assignment and conflict resolution are needed. A simple glossary of LCS terms and components follows.

LCS Terms and Components:

- Conditions – The circumstances under which a rule becomes active. Rules whose conditions match the state of the environment provided by the detectors are said to have fired.
- Actions – The method the LCS uses to affect its environmental state.
- Detector – The sensor interpreter that determines the environmental state to match against the rule's conditions.
- Effectors – Effectors interact with the environment according to the selected action.
- Rule Set – The set of all rules within the LCS, it can be thought of as the sum total of all knowledge the LCS has learned through interacting with its environment.

- Match Set – The set of all rules whose conditions match the current values of the detectors.
- Action Set – The subset of rules from the match set that is selected for action.
- Action Selection Mechanism – The method used to select the action out of the rules in the match set.
- Credit Assignment Mechanism – The facility that updates the rules' merit ratings based on feedback from the environment.
- Conflict Resolution Mechanism – When conflicting actions of equal merit are presented, this mechanism breaks the tie.
- Rule Discovery Mechanism – Method for discovering new (better) rules for the LCS.

A GA is used due to its global search properties and theoretically backed efficiency for processing (schema) information. Other methods, such as genetic programming (GP) have been used (Koza, 1992; Tufts, 1995). The possibility exists for other evolutionary methods to be used, although no other evolutionary methods have appeared in the LCS literature. The evolutionary approach to rule discovery is considered by some researchers to be the defining element of the LCS.

LCS Operational Cycle

The LCS starts by encoding its perceived state in the environment via its detectors. The detectors encode the state of the environment for the rules to be matched against. The matching rules form the match-set. Action selection takes the match-set and determines the winning action. Basing selection on a proportion of an action's cumulative merit, a stochastic action selection method may be used. Such a method was presented in Equation 16.1. The effectors perform the selected action. Credit assignment takes any available feedback from the environment and updates the merit ratings of the participating rules. The GA takes the rule-set(s) and modifies the rules by deleting inferior rules (or rule-sets). The GA also spawns new rules from highly fit rules, or rule-sets. GA activation may be periodic or triggered.[1] This process of natural selection is based on accumulated knowledge (fitness information) about the overall performance (accuracy) of the rules or rule-set. The process then repeats.

Credit Assignment

Credit assignment refers to the tuning of the existing rules' merit. Originally, this metric was termed the *strength* of the rule. Strength denotes a rule's predicted payoff for performing its action. Newer measures, based in statistical

---

[1] Periodic GA operation is the simplest form. If the system has a way to signal the GA to search for new rules then the GA may be triggered anytime the system generates the signal.

variance, suggest that direct payoff-based methods may lead to erroneous behavior, since the LCS may need to take a short-term sub-optimal action in order to obtain long-term optimality. The strength-based approaches sometimes become "greedy," i.e., the rule opts for the highest predicted payoff at the next time step.

Rule updates are commonly done via a *reinforcement learning* (RL) algorithm. RL is a class of problems that provides an update based on feedback from the environment, but not optimal (target) actions. Update algorithms such as Holland's *Bucket-Brigade* (Holland et al., 1986) and Watkins' *Q-Learning* (Watkins, 1989; Watkins & Dayan, 1992) are common methods for updating a rule's merit. Rummery and Niranjan (1994) provide an analysis of several of RL algorithms, including Q-learning and Bucket-Brigade algorithms. Their work suggests that in the long run (infinite horizon), both algorithms effectively perform the same update.

How does the GA come into play?

Because the LCS has a variable architecture, a modification method is needed to augment its rule-set. Genetic algorithms (GAs) may be used as global search mechanisms (Goldberg, 1989). In the context of the LCS, the GA is a *rule-discovery mechanism*.

The type of GA a LCS uses varies according the philosophy of rule search. There are "two camps" of thought for rule discovery in the LCS. These "camps" define the methodology for applying the GA in the LCS. The two approaches are named after the institutions that played a key role in their development. The *Michigan approach* was developed at the University of Michigan (Holland, 1978; Holland et al., 1986). The *Pitt approach* emerged from the University of Pittsburgh (Smith, 1980; Smith, 1983).

The Michigan Approach

The basic philosophy of the Michigan approach is to treat each rule in the rule-set as an individual in a population. Individuals compete via fitness for reproductive rights. Highly fit individuals have the opportunity to mate with other highly fit individuals thereby increasing the chance that their progeny will have better survival characteristics. This method of rule-discovery can be said to be a "population of rules approach," i.e., only one rule-set exists.

The type of GA operating in a Michigan-style LCS is typically a *steady-state GA* (DeJong, 1975; DeJong & Sarma, 1993; Whitley 1989). This form of GA selects a subset of individuals from the population. This subset is then subjected to recombination and mutation. Finally, the offspring of the subset replace existing individuals in the population.

The Pitt Approach

The Pitt approach chooses to view an *entire rule-set* as an individual with the population made up of multiple rule-sets. Using this approach, rule-sets compete to mate with other rule-sets; thereby, exchanging genes (rules). This method hopes to combine rules (genes) over many generations to form an effective rule-set. The Pitt approach can be thought of as a "population of rule-sets approach," i.e., many competing rule-sets exist. The type of GA used for the Pitt Approach may be either a *steady-state GA* or a simple GA.

Which Approach is Better?

There is no clear evidence that either method is superior to the other. The deciding factor relies on the environment (problem) that the LCS is immersed. A rule of thumb might be that for computationally intense environments, where the rules and rule-set may be quite complex, a Michigan approach might be appropriate since it is the one rule-set approach. For environments, where a large number of simple rules is expected, a Pitt approach might provide better search properties.

A Word on Message Lists

The *message list* component is sometimes found in LCSs. Message lists provide a way to temporally delay messages for internal message passing, i.e., rules signal other rules. This means that rule conditions do not necessarily match *environmental conditions* (as in the previous discussion), instead the rules may match the *internal state* as well. The message list acts as an interface between the detectors, effectors, and rules. Some problems involve *temporal delay*, which mandates some form of input queuing system (memory) that stores past inputs, or rule firings, in order to utilize that information in determining future actions.

The message list represents one of the most complex structures in the LCS. This complexity stems from a desire to have the system form *chains*, or sequences of rule firings, that lead to an eventual reward. The complexity of chain formation is quite daunting. Smith (1991) analyzed memory exploitation in LCSs with message lists. Smith studied the effects of internal message passing. His basic setup was to discover rules that trigger other rules in a learning automata framework. He found that *parasitic rules* inherently formed, which disrupted proper (effective) chain formation.

Disruption is best described by example. Consider a learning system, like an LCS, that discovers rules that lead to some form of reward. These rules by the original, economics-based strengths would grow in prominence. This is due to the system acting appropriately and receiving reward. To continually refine the rule-base, the GA is called to discover better rules. In the process of discovery, parasite rules are formed, which call (activate) the good rules, thereby receiving partial reward for participating in a chain of rule activations leading to a reward. The system could learn the problem, but with subsequent refinements, the GA would form internal messaging rules that, in effect, lie about the current state of the system just to get rewarded. The reason for this is the way reward (credit assignment) was performed. Formation of chains relies on the propagation of rewards received back through the message list. Rules that lead to the activation of other rules, which subsequently lead to reward, received credit for participation via the bucket brigade (or similar) algorithms. The parasitic rules eventually become so greedy, and aggressive that they trigger rules at inappropriate times. The parasite effectively lies about the state of the environment, or system, in hopes of receiving reward through the bucket-brigade. This improper triggering of actions eventually occurs at inappropriate times thus producing counterproductive behaviors.

The Current State of the Art

There is no "standard LCS." Some notable LCS implementations include the research of Wilson (1986; 1994; 1995). Wilson's work centers on the usage of LCSs in the simulation of adaptive behavior. His most current efforts have been with an advanced LCS called XCS (Wilson, 1995). For a more introductory system, Wilson provides ZCS, a *zeroth-level* LCS, which has the basic parts listed in this tutorial, excluding a message list.

The work of Smith and Cribbs (1994) relates artificial neural networks (ANNs) to the LCS. While ANNs may not be exactly the same as LCSs, there is a considerable amount of information that may be applicable to LCS research.

An initial framework by Valenzuela-Rendon (1991) introduces the principles of fuzzy logic to the LCS. More recently, the work by Bonarini (1996) is extending the fuzzy LCS concept. Fuzzy logic, like ANNs, may provide a rich medium of cross-talk to inspire future LCS research.

Finally, there is a variant of the GA that proves quite promising in the LCS. That is *genetic programming* (GP). GP is a tree based evolutionary algorithm that evolves program trees. This is a variable length structure that may allow for complex relationships to be mapped. This is extremely useful in the area of LCSs, where the system must learn from interacting with the environment. Little work has been done, but the advantages are quite attractive. Tufts (1995) has developed a GP-based LCS. Wilson's ZCS paper mentions that lisp *s-expressions,* which are similar to the lisp-like trees found in GP, might provide a compatible representation for LCS rules (Wilson, 1994).

ARTIFICIAL NEURAL NETWORKS AND THE LCS

Artificial neural networks (ANNs) provide a method to illustrate many of the ideas embodied by the LCS. This study shows a simple ANN application that involves classification of a series of binary patterns to a single binary signal. Linearly inseparable Boolean problems have become a staple of ANN benchmarks. These results stem from Smith and Cribbs (1994).

Input connectivity selection has been shown to extend the perceptron (Wilson, 1990). The addition of input selection provides a method to nonlinearly partition inputs. Using the premise of input partitioning, it becomes clear that LCS rules partition the input space through use of the *don't care* operator (#), while ANN's ignore inputs through zero valued weights (see Figure 16.2). The LCS uses 0, 1, and # to denote what values the conditions must meet for the rule to fire. A simple transformation of this scheme to -1, +1, and 0 easily emulates similar traits for the ANN (Smith & Cribbs, 1994). Figure 16.2 shows the overall LCS-like ANN. The ANN structure, presented in Figure 16.3, contains a hidden layer of nonlinear (bipolar sigmoid) activation functions and single output node (linear sum function). The update of weights is constrained to the hidden and output layer weights only.

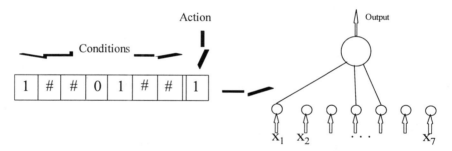

Figure 16.2. Mapping of an LCS rule onto an ANN computational unit (neuron).

In an LCS, the GA is responsible for rule discovery. The same premise can be used to evolve ANNs. Assuming a Michigan-style LCS model, the hidden layer of a multi-layer ANN may be considered the ANN's rule-set. The input to hidden layer connectivity of an ANN may be treated as the conditions for the individual rules (neurons). Fitness governs which rules, or neurons, are selected for evolution. Many fitness measures could be used, but effective measures appear to be those based in correlating the error (reward) signal to the active rules (or neurons). Fahlman (1990) suggests a similar form of correlation measure to grow "cascade-correlation" networks.

Equation 16.2 shows the fitness function used in an LCS-inspired ANN (Smith & Cribbs, 1994; Cribbs, 1995).

$$f_i = f_i + \sigma(x) + \overline{E} + [\sigma(x) \cdot \overline{E}] \cdot |w_{i,j}| \qquad (16.2)$$

The equation is shown in update fashion. The function σ denotes the activation function for the hidden layer neurons, and E complement (the bar above E) denotes when the ANN's training case (mean square) error is below a given threshold. This "digital" error threshold governs what the GA considers as the error state of the ANN. As long at the ANN's error is *below* this value, E is 0-- indicating a "no error" state. Using this logical format, the rest of the equation acts as a "bonus" function that provides additional fitness point for firing, when the error threshold has not been exceeded. The multiplication by the absolute value of $w_{i,j}$ is the magnitude of the output weight of hidden neuron, i to output node j. The multiplier promotes bonuses that are not "parasitic." Parasites, in this sense, are rules (or neurons) that act similarly to other rules active at the same time, but do not additively participate, i.e., parasites do not affect the result appropriately.

A typical GA attempts to evolve the individuals in a population until the population consists largely of one type of individual. This form of evolution does not apply to the Michigan style LCS, or ANNs, because a *diverse collection* of rules must be maintained to allow the system to interact efficiently with its environment. To accomplish this task, several methods to preserve the diversity within a population have been developed. Goldberg (1989) developed a computational method called *explicit sharing*. The premise of sharing is to measure the "distance," in the genotype space, of each individual with respect to the rest of the individuals in the population. The fitness values for each individual are degraded based on their proximity to the other members of the population.

Explicit sharing is difficult to implement and computationally expensive. A simpler method is *implicit sharing* (Horn et al., 1994; Smith, 1993). This form of sharing is based on information about the way the individuals are interpreted in phenotype space, i.e., the similar behaviors between individuals indicate similarity. For the purposes of this study, implicit sharing is easier to implement and well suited to ANNs.

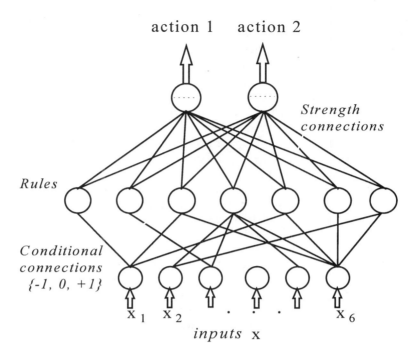

Figure 16.3. A complete LCS-like ANN. Notice the conditional connections, they implement the input to the rules (hidden layer neurons). The strength connections denote the strength of a given rule with respect to an action.

Using Equation 16.2, the "raw" fitnesses are calculated. To implement implicit sharing, the state information about each neuron (whether it is active or not) is recorded. This information is then used to degrade the fitnesses by the number of active and inactive nodes. In addition, a constraint of unique connectivity is imposed on the population. This is done because redundant rules are unnecessary. GA populations, typically, have multiple copies of an individual (many individuals of the same specie), to compensate a *copy count* is kept to emulate multiple copies of a specie. The concept of copy count and unique rules can be found in Wilson's (1994; 1995) work on LCSs.

$$S_i = \frac{1}{\sum_{\forall j \in H} (\sigma_i \wedge \sigma_j)} + \frac{1}{N_i} \qquad (16.3)$$

Equation 16.3 shows the sharing method applied to the fitness calculated by Equation 16.2. The domain of this function is every node in the hidden layer, *H*. The first term is sharing between like acting nodes, i.e., it compares node i's firing state to node j's and returns 1 for a match and 0 otherwise. For the purposes of this study, a node's output will either be +1 or -1 (the latter indicating the "active" state). The first term can be considered sharing by like acting neurons (sharing by behavior or phenotype), while the second term simply shares by the number of copies (sharing by specie or genotype) that "exist in the population." Sharing is applied by multiplying the fitness value of an individual by the corresponding sharing value.

The novel LCS approach to ANNs appeared first as a *batch learning* method that evaluated each input pattern of a training set and formed fitness values and sharing terms only after all the patterns in the training set had been evaluated (Smith & Cribbs, 1994). The test problem used to validate the "batch version" is called the "multiplexor problem" due to the similarity of function with the discrete electronic device of the same name. The multiplexor problem description and results of an *incremental learning* version are presented in the next section. Equations 16.2 and 16.3 represent the "incremental versions" used to obtain the results presented.

THE MULTIPLEXOR PROBLEM

The multiplexor problem was shown by Wilson (1990) to be an effective testbed for GA-based input partitioning. A stochastic (incremental) version of this problem was used to test the characteristics of online learning in an ANN (Cribbs, 1995). The selected test problem consists of 64 input patterns, mapping 1 of 4 input signals (either -1 or 1) based on two address lines. The number of inputs to the ANN is the 2 address lines plus the 4 input signals. The goal of the problem is to select the correct output signal (either -1 or 1).

Figure 16.4 shows the performance history over a training run. The figure shows the ANN error sampled at 128 training cycles per data point. Each data point represents an exponentially smoothed "running average" of the system's performance (training case error). The training parameters used to obtain this result correspond to the values in Table 16.1.

While its performance is worse than Cribbs (1995) incremental result, the system is able to learn a mapping for the problem. A discussion of the ANN's performance on the 6 multiplexor test problem follows.

Discussion

The performance history (Figure 16.4) shows that an ANN with genetically evolved input connectivity can classify the correct values for the input parameters. Furthermore, the results indicate that the difference in sharing function contribute to the overall final network.

It was pointed out earlier that sharing was responsible for speciation via degrading fitness. Degradation of fitness based on the composition of the population led to an interesting difference in speciation. Varying levels of speciation using implicit sharing are possible by inclusion or exclusion of terms in the sharing equation. By direct sharing of fitness by the number of a given specie, uniqueness becomes a deciding factor in the size of the population. Figures 16.5 and 16.6 depict the evolutionary history of a training run on the 6 multiplexor with sharing by copy count and sharing by firing behavior (see Equation 16.3). Using the copy count of a specie generates a larger ending population. Lower population sizes (lower levels of speciation) were observed when sharing utilized only the first term of Equation 16.3, i.e., sharing type firing behavior only (see Figures 16.7 and 16.8).

Another point about the results presented involves the output activation function. The structure of the ANN was a hidden layer of *bipolar sigmoidal* activation functions[2] and a linear weighted (sum) output (linear activation function). Additionally, training of the ANN involved only the hidden to output weights. The two constraints, linear weighted sum and one layer of modifiable weights, lead to an approximator that can only update its weight via gradient descent. All nonlinear effects emerged from the GA evolved input to hidden connectivity. The LCS paradigm uses GA evolved rule conditions, much like the above ANN. This similarity leads to the aforementioned analogy of an LCS as a type of ANN (Smith & Cribbs, 1994).

Differences in the LCS and ANN prevail in the interpretation of rule (hidden layer neuron) firing. The ANN simply takes a *weighted sum* of all the active neurons, while the LCS sums all active rules with the same action. This summing effect in both cases is similar, but the difference becomes apparent considering *negative valued weights*. Negative values in an ANN act as analog complements, i.e., the negative value weights oppose an action. Opposition in the LCS takes the form of competing rules advocating different actions. The cumulative effect is that the LCS relies on the GA to "fill in the blanks," i.e., discover rules to cover conditions when the existing rule set does not perform in a desired fashion. The ANN takes another approach; allowing the gradient descent procedure to change the interpretation of the neuron activations.

---

[2] Bipolar activation denotes binary output centered around zero. The active state is positive one (+1) and the rest, or "off," state is negative one (-1). Sigmoidal activation indicates continuous activation with saturation occurring quickly at one of the two extremes (-1 or +1).

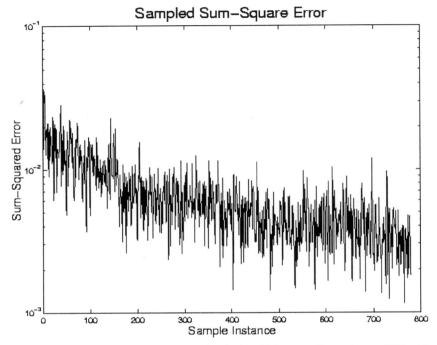

Figure 16.4. Performance (learning) history for the 6 multiplexor problem using an ANN with GA-based input partitioning. Note that this result uses the fitness and sharing scheme defined by Equations 2 and 3.

Table 16.1. Training parameters used the train the ANN for the 6 multiplexor problem.

Parameter	*Value*
Learning Rate, η	0.02
Exponential Smoothing Constant[3], φ	0.10
"Digital" Error Threshold	0.05
GA Trigger Frequency	every 675 epochs
Selection Block Size	2 individuals
Probability of Recombination, p_c	0.90
Probability of Mutation, p_m	0.01
Probability of Hash Mutation, p_h	0.10

---

[3] Exponential smoothing is defined as a continuous averaging method. The method may be defined as,

$$S^{t+1} = (1 - \varphi) \cdot S^t + \varphi \cdot \hat{s}$$

where "s hat" is the current measurement and S is the running average.

The ANN's gradient descent mechanism can change the output weights of all active neurons; hence, the gradient descent mechanism continually tunes the existing neurons' weights. Continual tuning could potentially modify an advocate neuron into an opponent if environmental cues exist to force a weight change. The results presented are not quite as good as those presented by Cribbs (1995), but the linear output of the ANN in this study contributes much to this fact. Cribbs (1995) work entailed bipolar sigmoid activation functions throughout the ANN (including output node). The added nonlinear activation at output confers greater classification ability than the linear activation used on output in this study.

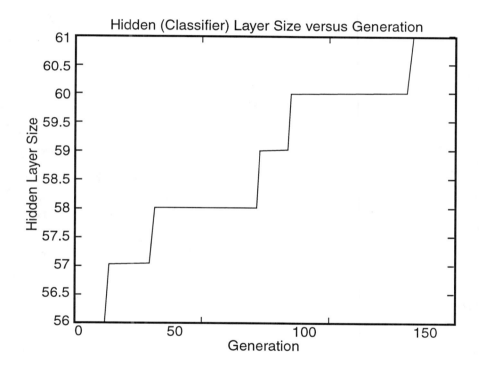

Figure 16.5 – Generational history of population (hidden layer) size. Note this result reflects sharing by both firing behavior (phenotype) and by specie count (genotype).

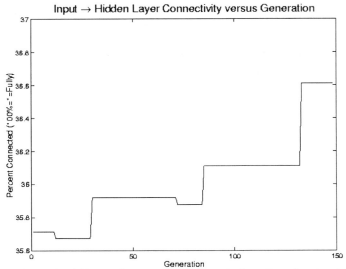

Figure 16.6. Generational history of connectivity between the input lines (layer) and the hidden layer (by percentage, 100% indicates full connectivity).

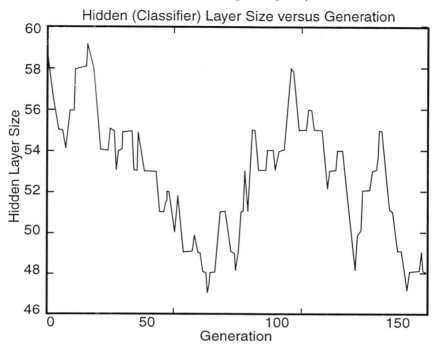

Figure 16.7. Generational history of population (hidden layer) size. Note this history depicts evolution with sharing by phenotype only, i.e., no sharing by specie (genotype).

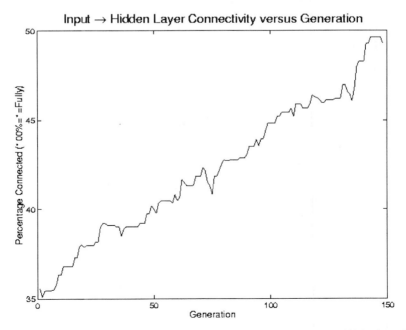

Figure 16.8. Generational history of connectivity between the input lines and the hidden layer (by percentage). Note this result emerged from sharing by firing behavior only.

Conclusion

This chapter presents a discussion of the major components of the LCS paradigm (including references to help the reader discover more about the LCS), a review of an analogy of LCS principles in a widely known machine learning technique, i.e., ANNs, and an example where GA-induced nonlinearity via input partitions contributed to the solution of a linearly inseparable problem.[4]

The LCS represents an interesting facet of machine learning. Genetics-based machine learning techniques take advantage of evolutionary search as their main search mechanism. Other, more traditional computational techniques are prevalent in machine learning with the GA-based avenues less explored. Finally, the presentation of *behavior-based sharing* shows a method of speciation compatible with ANNs.

ACKNOWLEDGMENTS

The authors wish to acknowledge the financial support from the National Science Foundation (grant ECS-9212066) and from NASA's Graduate Student Researchers Program (contract NGT 4-52403).

The authors thank the editors for their invitation to be included in this project. Much insight was gained during the GA course at The University of Alabama and they have provided a forum to showcase the students' projects.

Lastly, the first author acknowledges his family's great support while staying in school longer than most students. Without their support, life would be "not so grand" and graduate school would probably have taken its toll long ago.

REFERENCES

[1] Bonarini, A. (1996) Evolutionary learning of fuzzy rules: Competition and cooperation. In W. Pedrycz, ed., *Fuzzy Modeling: Paradigms and Practice*, Kluwer Academic Press, Norwell, MA, pp. 265-284.

[2] Cribbs, H.B. (1995) Cooperative learning classifier systems: A neural network approach. Master Thesis (TCGA Report No. 95002). The University of Alabama: Tuscaloosa, AL.

---

[4] As previously pointed out, the use of a linear activation function on the output shows that the nonlinearity derived from input partitioning confirms Wilson's (1990) result.

[3] Cybenko, G. (1989) Approximation by superpositions of a sigmoidal function. *Mathematics of Control, Signals & Systems*, 5, 45-67.

[4] DeJong, K.A. (1975) An analysis of the behavior of a class of genetic adaptive systems. Ph.D. Thesis, University of Michigan, Ann Arbor.

[5] DeJong, K.A. & Sarma, J. (1993) Generation gaps revisited. In L.D. Whitley, ed., *Foundations of Genetic Algorithms* 2. Morgan Kaufmann: San Mateo, CA.

[6] Fahlman, S.E. & Lebiere, C. (1990) The cascade-correlation architecture. Technical Report (CMU-CS-90-100). Carnegie Mellon University: Pittsburgh, PA.

[7] Goldberg, D.E. (1989) *Genetic algorithms in search, optimization and machine learning*. Addison-Wesley: Reading, MA.

[8] Holland, J. H. & Reitmann, J. (1978) Cognitive systems based on adaptive algorithms. *Pattern-Directed Inference Systems.* Academic Press, Inc.: New York.

[9] Holland, J. H., Holyoak, K. J., Nisbett, R. E., & Thagard, P. R. (1986) *Induction: Processes of inference, learning, and discovery.* Cambridge: MIT Press.

[10] Horn, J., Goldberg, D.E., & Deb, K. (1994) Implicit niching in a learning classifier system: Nature's way. *Evolutionary Computation, (2)* 1, pp. 37-66.

[11] Hornik, K., Stinchombe, M., & White, H. (1989) Multilayer feedforward networks are universal approximators. *Neural Networks,* vol. 2, pp. 359-366.

[12] Koza, J. (1992) *Genetic Programming: On programming computers by the means of natural selection.* MIT Press: Cambridge, MA.

[13] Minsky, M. & Papert, S. (1969) *Perceptrons: An introduction to computational geometry.* MIT Press: Cambridge, MA.

[14] Rosenblatt, F. (1958) The Perceptron: A probabilistic model for information storage and organization in the brain. *Psychological Review* 65, pp. 386-408.

[15] Rumelhart, D. E., Hinton, G. E., & Williams. R. J. (1986) Learning representation by backpropagating errors. *Nature, 323,* pp. 533-536.

[16] Rummery, G. A. & Niranjan, M. (1994) On-line Q-learning using connectionist systems. Cambridge University Engineering Department (Report No. CUED/F-INFENG/TR 166). Cambridge CB2 1PZ, England.

[17] Smith, S.F. (1980) A learning system based on genetic adaptive algorithms. Unpublished Doctoral Dissertation. University of Pittsburgh.

[18] Smith, S.F. (1983) Flexible learning of problem solving heuristics through adaptive search. *Proceedings of the 8th International Joint Conference on Artificial Intelligence*, pp. 422-425.

[19] Smith, R.E. (1991) Default hierarchy formation and memory exploitation in learning classifier systems. Ph.D. Dissertation, The University of Alabama: Tuscaloosa.

[20] Smith, R.E., Forrest, S. & Perelson, A.S. (1993) Searching for diverse cooperative populations with genetic algorithms. *Evolutionary Computation,* (1)3, pp. 127-149.

[21] Smith, R.E. & Cribbs, H.B. (1994) Is a learning classifier system a type of neural network? *Evolutionary Computation,* (2) 1, pp. 19-36.

[22] Smith, R.E. & Cribbs, H.B. (1996) Cooperative versus competitive system elements in coevolutionary systems. *FROM ANIMALS TO ANIMATS 4: Proceedings of the 4th International Conference on Simulation of Adaptive Behavior.*

[23] Tufts, P. (1995) Dynamic classifiers: Genetic programming and classifier systems. In *AAAI Fall Symposium.* AAAI.

[24] Valenzuela-Rendon, M. (1991) The Fuzzy classifier system: A classifier system for continuously varying variables. *Proceedings of the Fourth International Conference on Genetic Algorithms*, pp. 346-353.

[25] Watkins, J. C. H. (1989) Learning with delayed rewards. Unpublished doctoral dissertation. King's College, London.

[26] Watkins, J.C.H. & Dayan, P. (1992) Technical Note: Q-learning. *Machine Learning* **8**, pp. 279-292.

[27] Whitley, D. (1989) Analysis of the GENITOR Algorithm and selection pressure: Why rank-based allocation of reproductive trials is best. *Proceedings of the Third International Conference on Genetic Algorithms.* Morgan Kaufmann: San Mateo, CA.

[28] Wilson, S.W. (1986) Classifier systems and the animat problem. The Rowland Institute of Science (Research Memo RIS No. 36r). Cambridge, MA.

[29] Wilson, S. W. (1990) Perceptron Redux: Emergence of structure. In S. Forrest (ed.) *Emergent Computation: Proceedings of the Ninth Annual International Conference of the Center for Nonlinear Studies on Self-Organization, Collective,   and Cooperative Phenomena in Natural and Artificial Computing Networks.* pp. 249-256. North-Holland: Amsterdam.

[30] Wilson, S.W. (1994) ZCS: A zeroth level classifier system. *Evolutionary Computation,* (2) 1, pp. 1-18.

[31] Wilson, S.W. (1995) Classifier fitness based on accuracy. *Evolutionary Computation,* (3) 2, pp. 149-175.

CHAPTER 17

GENETIC ALGORITHMS FOR GAME PLAYING

Jiefu Shi
shij@agcs.com

ABSTRACT

This chapter examines genetic algorithms (GA) and machine learning using the game of tic-tac-toe. After "learning" acceptable strategies for playing the game, the GA-driven computer player is able to play a competent game of tic-tac-toe. Results obtained using a GA are compared to results obtained using alternative AI techniques.

INTRODUCTION

Game playing has long been a research favorite of the AI community. AI researchers have developed numerous search methods for game playing. The search methods developed can either be brute force search or a more "intelligent" heuristic search. The current project focuses on the simple game of "Tic-Tac-Toe." Genetic algorithms are used to develop strategies under different situations, and through a "survival of the fittest" rule, eliminate bad strategies and allow the good strategies to survive.

Even though "Tic-Tac-Toe" is a very simple game, the goal of the current project is to explore game playing strategies using genetic algorithms. Because of the robustness of GAs, the parameters and operators chosen to solve this problem should be applicable to other similar problems in the realm of game playing.

LITERATURE REVIEW

There are many books and papers on game playing, machine learning, and genetic algorithms. The following list contains a few which are related to this problem.

One of the first papers written on game playing and machine learning is *Some Studies in Machine Learning Using the Game of Checkers* (Samuel, A. L., 1959). In this paper, Samuel examined machine learning and the game of checkers. Samuel developed 26 parameters which he used to determine how "good" a particular board position is. Each parameter has a value associated with it to specify its relative importance. He constructed a polynomial function

with these parameters to give each position a value. The polynomial looks like:
position value $= p_1x_1 + p_2x_2 + p_3x_3 + ...$

His goal was to find the x_n values that represent the most effective strategy for playing checkers. He started out by assigning the x_n random values and let two strategies play each other, kept the set of values which won the most games, and continued to generate new values. After many games were played, he found that the strategy which survived at the end had effectively "learned" to play the game of checkers.

Unfortunately, as in most situations, this random search can take quite a long time. However, Samuel did produce a good strategy which was able to beat the best human players of his day. The current project is similar to the work of Samuel. Here, the hope is that a search conducted using a GA will allow for the development of more effective strategies in less time.

Another early work on game playing was conducted by Michie (1962). In this paper, Michie examined the process of machine learning and the game of tic-tac-toe. He found around 300 distinct positions that the opening player can be faced with. He then took all the possible moves at each position and assigned a value to represent its relative merit; the higher the value, the better the move. He used a selection process that chose the move that was associated with the highest value. If the move resulted in a win, he increased the value associated with that move. If the move resulted in a loss, he decreased the value. So at first, the game was played randomly. But as the game playing program won and lost games, it learned a little more about the game. After a number of games played, the machine was successfully trained to play the game of tic-tac-toe.

Here, a similar approach is employed. However, in the current effort a GA is used to generate new strategies. These strategies are encoded into strings, and there exist a population pool of more than just one strategy to be considered at each iteration. Michie used his machine to play human players. Here, the GA-generated strategies are pitted in competition against other computer strategies.

A good introductory textbook on AI is Ginsberg's *Essentials of Artificial Intelligence* (Ginsberg, 1993). In his book, Ginsberg outlined the different search algorithms and heuristics which are used in the AI community to solve game playing problems. To solve the current problem, the game tree is expanded and a search algorithm is used to traverse the tree in hopes of finding an optimal solution each time a move is to be made. Here, enumeration would result in an optimum solution, but it is time consuming to the point of infeasibility.

There have also been several successful attempts at machine learning using neural networks. A good reference on neural networks and machine earning is *Neural Networks - A Comprehensive Foundation* (Haykin, 1994). In his book, Haykin discusses neural networks and the different types of learning including supervised learning, unsupervised learning, competitive learning, and others. Certainly, the neural network approach is feasible in a variety of problems (perhaps even here). A neural network tic-tac-toe program would be an outstanding opponent for the GA tic-tac-toe game playing program. Unfortunately, such a comparison is not presented in the current chapter.

The textbook *Genetic Algorithms in Search, Optimization & Machine Learning* (Goldberg, 1989) serves as a good introduction to GAs. Goldberg covers the basics of GAs along with providing a PASCAL computer code for implementation. Goldberg's code proved to be very helpful for the current project. Even though the program generated in this effort was written in C, the PASCAL version was used as a reference to the standard GA operators such as crossover and mutation. Goldberg has a rather large section in his book devoted to machine learning via classifier systems. This section of his book also provided several good ideas for the current problem.

An early attempt at machine learning using GAs was done by Bagley on the game of hexapawn. In the paper, *The behavior of adaptive systems which employ genetic and correlation algorithms* (Bagley, 1967), Bagley examined GA and game playing. He used the game of hexapawn, which is played on a 3 X 3 chessboard with three pawns to each side. He used the basic reproduction, crossover, and mutation operators as well as diploid strings, dominance, and inversion operators. His program performed well after running through a number of generations.

Tic-tac-toe, however, is more complicated than the hexapawn game. Whereas hexapawn has only 52 unique configurations, tic-tac-toe has 612. This makes the current problem much more difficult to solve since there are far more points in the search space.

PROBLEM DESCRIPTION

The Game

The game of "Tic-Tac-Toe" is quite simple; generally thought of as a "kid's game." It is played on a 3 x 3 board. Initially, the board contains nothing but blank squares. The two players, X and O, each place X's and O's on the board, one at a time, alternately. The player who places three of his pieces in a row either horizontally, vertically, or diagonally wins the game. The game is a tie if nobody has won after the board is filled.

Traditional AI Methods

Traditionally, there are two classes of methods for solving a problem such as playing Tic-Tac-Toe: (1) the brute force method or (2) some sort of heuristic search method.

In solving such a search problem, the decision tree can be searched breadth first, expanding the game playing tree as far as possible, and examining each possible move and its merits. By looking ahead in this fashion, the ramification of each possible move can be examined using a minimax technique which involves computing a function value to evaluate possible solutions and assign a numerical value to each move according to player's chance of winning the game. The higher the probability of winning the game as a result of making a particular move, the higher the evaluation function value assigned to that particular move. A computer program employing such a strategy would then select the move that results in the maximization of its own score while simultaneously minimizing its opponent's score.

The problem with the aforementioned approach is that for many games, including tic-tac-toe, the search tree grows too big too quickly. The combinatorial explosion prohibits searching far enough down the decision tree for effective solutions.

As an alternative, any of a number of heuristic search rules can be used to eliminate branches of the search tree from consideration, to prune the tree. This is in effect a depth first search. Such an approach requires a lot of effort on the programmer's part to carefully choose which branches to discard. For most problems, however, the breadth first search method is still used even though it is much less efficient.

A problem shared by both the breadth first and the depth first methods is the lack of robustness. Once a program is written to play a specific game, it requires a lot of work to change the evaluation function and the search methods to play a different game, even though it might be a slight variation of the original game. Thus, such a program is only as good as the evaluation function, and it does not learn from its mistakes to get any better.

GA and Game Playing

Genetic algorithms are search algorithms based on the principle of natural selection. For game playing, each string in a GA population can represent different strategies for playing the game. Initially, a fixed number of strings are generated randomly. Then, during the selection process, each strategy is required to play a certain number of games against other strategies. A win-loss record is kept for each strategy. The ones that perform badly are discarded while the strategies with the most wins are kept. After selection, crossover, and mutation are performed on those strings to create new strings which form the new population. After this process is repeated a number of times, the population pool will consist of strategies that are fit to survive (i.e., playing a good game of tic-tac-toe).

GA IMPLEMENTATION

Configuration Encoding

To develop a strategy for the game of tic-tac-toe, there exists a desired move for each board configuration. The game of tic-tac-toe is played on a board of nine squares. Each square can either be marked by an "X," an "O," or a blank. The total number of possible configurations is 3^9 or 19683. Here, it is assumed that a ternary string of length nine is used to code each possible configuration. "0" is used to represent a blank, "1" is used to represent an "X," and "2" is used to represent an "O." For example, the following board could be represented as shown:

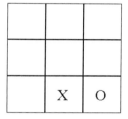

Ternary Number = 000000012
Decimal Equivalent = 5

Initialization

First, all configurations possible within the game of tic-tac-toe must be determined. As this is obviously a time-consuming task, a computer program was written to find all valid configurations. For the "X" player, which is the player that goes first, the number of Xs and Os on the board at that time must be equal. Conversely, for the "O" player, there must be exactly one more X than the number of Os on the board. There are also a lot of configurations that are equivalent to each other. For example, there are configurations that have mirror images or transposes. These equivalent configurations must be determined to keep the length of the GA strings as short as possible.

A computer program was written to find all possible configurations for the "X" player. These configurations were saved to a file. The file looks like the following table:

Config.	0	5	7	11	15	19	...
Index	1	2	3	4	2	4	...
Variation	0	0	0	0	1	2	...

The index is the position in the string in which that particular configuration is stored. Notice that configurations 1-4 were not in the table because they are invalid for the "X" player. Configurations 5 and 15 contain the same index which means that 5 and 15 are variants of the same configuration:

Configuration 5:   Configuration 15:

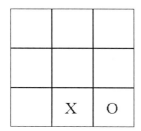   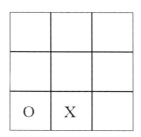

As can be seen from the above figure, configuration 15 is the same as configuration 5, it is just flipped horizontally. There is not need for an extra space in the string to represent 15. In the GA implementation, the table will be referred to in order to discern that configuration 15 is the same as configuration 2. This same situation exists for configuration 19; it is the same as configuration 11. Thus, both of them share the same index. The above table also has a "variation field" to depict how the configuration is related to the original. For example, let "0" denote that the configuration is the original, "1" denotes that the configuration is the original flipped horizontally, "2" denotes that it is the original flipped vertically, etc.

After the aforementioned program was executed, it returned 612 indices, which means that there are 612 unique configurations for the "X" player.

Such a file was also created for the "O" player, which contains all the valid configurations and their indices.

String Representations

Since there are 612 different configurations and each configuration can have nine responses, a string of length 612 is needed if the alphabet 0 to 8 is to be used. Each bit will represent a move corresponding to that configuration index. Alternatively, a ternary alphabet of 0, 1 and 2 could be used, but this would double the length of the string to 1224.

Using the first method:

Index	1	2	3	4	5	6	...
Content	5	0	4	6	3	8	...

Using the second method:

Index	1	3	5	7	9	11	...
Content	12	00	10	20	10	21	...

Both strings represent the same strategy. In general, the GA works better with a smaller alphabet. Therefore, method number two is used here.

String Decoding

To discuss the coding of the GA strings, consider the configuration table listed above. If, as the "X" player, the following board configuration is encountered:

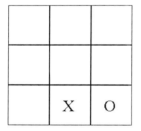

This board is encoded as $000000012_{(base3)} = 5_{(base\ 10)}$. By consulting the configuration table, the index of configuration 5 is determined to be 2. Next, the string at index 2 is used if the first method is being used, or the string at index 3 if the second method is being used. Since the second method is being used here, the result is 0. Thus, the mover for the first board is to place an "X" at board position 0. After "X" makes this move, the board will appear as follows:

If the string suggests making a move at a position that is already occupied, say position 8 for the above example, then a randomly selected valid move is made, and the string is altered so that this move is captured in the genetic material. Of course, if it is a good move, then that string will thrive in future generations, whereas, if it is an ineffective move, then said string will die out of the population.

When selecting a move, the variation field must also be consulted so as to make any translations necessary to reflect that variation.

String Length Reduction

Certainly, there is motivation for reducing the length of the strings on which a GA must operate; if nothing else, this reduces the size of the search space. In the current case, the strings could be as long as 1000 bits; this might not be such a good idea since the corresponding search space would contain 2^{1000} points or possible solutions to the problem at hand. Basically, it would be extremely difficult and time consuming for any search method to traverse this space and ensure a quality solution had been found. Thus, it is worthwhile to consider mechanisms of reducing the string length. One way to reduce the length is to reduce the number of decisions that the strings must represent. Instead of having the GA make every decision, the problem could be structured so that the simple decisions are not represented by the strings; domain knowledge is included. For instance, there are at least two predicaments for which any effective player will make the correct decision almost 100% of the time. First, if a player can win the game by placing its mark on one of the squares during the next move, then return that move. Second, if the opponent can win the game by placing its mark on one the squares during the next move, then return that move.

By using those rules, the string length can be reduced because it become unnecessary to store situations that fit the above description in the strings. By doing so, the number of configurations can be reduced down to 104. This means that the string length is 208 when a ternary alphabet is employed.

Fitness Function

Two ways of coding the fitness function are considered here. First, the "X" and the "O" players could be pitted against one another in a series of games. Or, alternatively, the game-playing program could be matched against a fixed algorithm. If the two players were forced to play one another, then two population pools would be generated initially. One pool would contain the "X" player strategy, the other pool would represent the "O" player strategy. In this approach, if 30 strings were generated in each pool, and each pool played five games, there would be 30*30*5 = 4500 games played total in each generation. The fitness function value of each string is derived from the win-loss-tie record of the strings.

On the other hand, if the strings are pitted against fixed algorithms, each strategy would play against its opponent a fixed number of games. Then, the fitness value would be determined in the same way as above; according to the win-loss-tie record. In this approach, if there are 100 strings in the pool, and each string is forced to play 50 games, there would be total of 5,000 games played during each generation.

The following scoring methods are used:

Player	Win	Tie	Loss
Player X	+4	+1	0
Player O	+5	+2	0

The difference between the "X" player and the "O" player is that for the "X" player, its objective is to win, because it goes first. It receives a smaller score for a tie because it has the advantage of moving first. But, as for the "O" player, because it goes second, its objective is to prevent the "X" player from winning. Thus, the "O" player is considered successful (to some extent) if it ties the game, and an even bigger accomplishment if it wins the game.

Using the above table, for example, if the "X" player has a record of 11-5-4 after 30 games, then it will have a fitness value of 49. If the "O" player has the same record, then it will have a fitness value of 65.

Crossover & Mutation

After the random generation and evaluation of the first generation, there exist 30 strings with associated score or fitness values. Next, roulette wheel selection is implemented to generate a mating pool of 30 strings. For example, if the initial generation was represented by the strings in the table below, then string number one would have the probability of 34/430 or 8% chance of making it to the next generation. Over the 30 selections, it would have a probability of 91% to have one or more copies of itself in the next generation. Likewise, string two would have a probability of 0.9% over one selection and a probability of 24% over all 30 selection.

String	1	2	3	4	5	...	Total
Score	34	4	23	12	2	...	430

The idea is, of course, to give the strings with a high fitness values a greater chance of making it to the next generation. And by using roulette wheel selection, it will add some variations into the next generation.

For this application, a crossover probability of 1.0 was used. This means that crossover will happen to all strings in the mating pool. A mutation probability of 0.001 was used.

Selecting GA Parameters

Initially, the GA was used to train the "X" player. A ternary alphabet was used consisting of 0, 1, and 2, so each move had two bits which combine to produce a number between 0 and 8. The GA strategy was developed by pitting it against a computer opponent that played with a very simple, perhaps "minimal" strategy (to be described below).

First, the board is initialized and place in a file so that the GA did not have to compute the boards every time it evaluated the fitness of a string. The board initialization program goes through the entire board configuration and saves the valid configurations while discarding the invalid ones. It also discards the boards which the "X" player can win during the next move and the ones "X" player must block in order to stay in the game. This is done because this domain knowledge was pre-programmed. For each valid configuration, the code applied each of the six variations and compared these variations of the current decision to the stored configuration. If the configuration is a variant of a previous one, it stores the current configuration, the original configuration, and the variation number. All the information is then saved to a data file.

After these preliminaries are completed, the GA program can now be executed. Each individual in the 100 string population is forced to play a computer opponent. This opponent uses following strategy:

1. Win the game: go through all the blank squares and, if it can win by placing its piece in one of them, go ahead and make that move.
2. Block opponent: go through all the blanks again and, if the opponent can win by placing its piece in one of them, then return that move.
3. Else, make a random move.

This, obviously, is not a very sophisticated strategy. However, this player does know when to block and when to win the game. This is enough to train the GA player with.   Here, the goal is to train the GA player so that it can beat this opponent virtually every time. If this is achieved, then the strategy employed by the GA is deemed to be effective for playing the game of "Tic-Tac-Toe".

Each string or individual plays 50 games against its simplistic computer opponent. Using the scoring method described above, a maximum of 200 points can be scored.

RESULTS

First, consider the on-line and off-line performance of the GA population. The on-line performance measures the average fitness of all individuals through the generations. The off-line performance is the average of the best individuals over generations. Since the GA is a stochastic method, it was run ten times to help dampen out any probabilistic errors. Figure 17.1 shows that the on-line performance starts out around 80, which is fairly weak (about 20 wins over 50 games). But then it starts improving rather quickly, up to 140 just after 30 generations, and increases all the way to 165 after 70 generations. After 100 generations, it is past 170 which is very good, considering the optimum is 200.

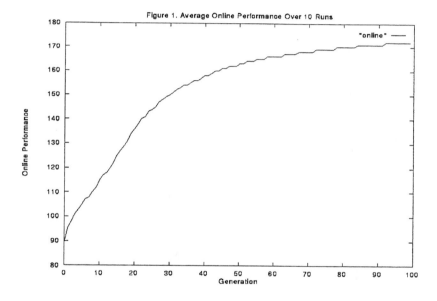

Figure 17.1 – On-line performance of the genetic algorithm.

Figure 17.2 is a plot of the off-line performance. The off-line performance starts out around 130, and begins to converge rather quickly. After 40 generations, it is around 180, and after 100 generations the off-line performance is almost 195. The best-ever string has a fitness value of 200, which means that it won every one of its games.

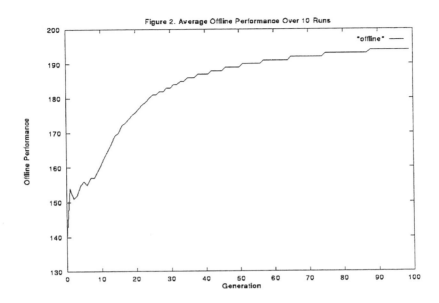

Figure 17.2 – Off-line performance of the genetic algorithm.

An issue that needs to be considered is random moves made by the GA. Recall that in the strings, an invalid move is replaced by a randomly generated
move. This random move is then put back in the string. Figure 17.3 shows the
percentage of moves that were made randomly over the generations. The GA
starts out by making almost 20% of the moves randomly, but as generations
progress, the percentage starts to drop. After 100 generations, the number of
moves made randomly is only around 3%. This indicates that as the GA begins
to play the games "better" (as shown on the performance graphs), the strings
also rely less and less on random moves. This is a good example of exploration
versus exploitation.

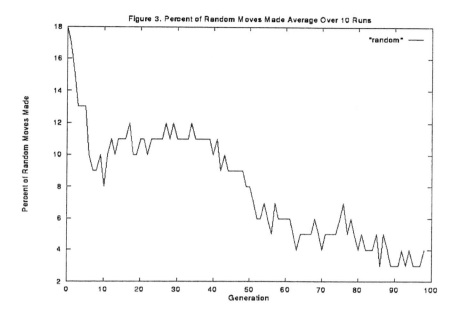

Figure 17.3 – Percentage of random moves made by the genetic algorithm.

The on-line and off-line performance plots might not tell the whole story. A few games were played with the strings in the GA that have top fitness values to see if they have really mastered the game as their fitness measures indicate. The results are documented in the next section.

Sample Games

<u>Game 1</u>

Computer Moves:

		X

O		X
		X

O		X
	X	X
		O

O		X
O	X	X
X		O

Human Moves:

O		X

O		X
		X
		O

O		X
O	X	X
		O

X wins the game. As one can see, X plays a very aggressive game (as the opening player should). X does not just make moves aimlessly, but it actually tries to set up its plays, the O player failed to make the right first move and X player took advantage of that.

Game 2

Computer Moves: Human Moves:

```
 .  .  X         .  O  X
 .  .  .         .  .  .
 .  .  .         .  .  .

 .  O  X         .  O  X
 .  .  .         .  .  O
 .  .  X         .  .  X

 .  O  X         O  .  X
 .  X  O         .  X  O
 .  .  X         .  .  X

 O  O  X
 .  X  O
 X  .  X
```

X wins the game. Again, X tried to set the human up and the human failed to make the right response for X's opening move. There was no way the X player could have been prevented from winning the game after the first "O" move was selected.

Game 3

Computer Moves:	Human Moves:

	X	

	O	
	X	

	O	X
	X	

	O	X
	X	
O		

	O	X
	X	X
O		

	O	X
O	X	X

	O	X
O	X	X
O		X

X wins the game. This time a different opening was used, but as before, the GA strategy indicates an attempt to set up its human counterpart.

Execution Performance

   One popular concern with GAs seems to be execution time.   When an algorithm is computing the effectiveness of 100 possible solutions in each generation, and then being run for numerous generations, the number of possible solutions evaluated can become significant.   Thus, it is important to consider the

amount of computational time associated with determining a solution. The GA code the current effort was run on is a SUN SPARCstation 20. On average, each generation took approximately 3.8 seconds, so over 100 generations, the program ran for just under 8 minutes.

CONCLUSION

Learning Classifier System

It turns out that the current effort is basically a simplified version of the "Pittsburgh" approach to the learning classifier system. The positions of the bits in the strings are the conditions. The actions are the contents of the strings. Here, all the possible states have been encoded into the strings. The selection process is basically a reward mechanism. Strings with good fitness values are allowed on to the next generation while the poorly fit strings are eliminated.

GA Versus Traditional AI Approach

In the early version of the GA program, the GA was allowed to make all of the decisions regarding a game, even the simple obvious moves. This resulted in longer strings, 618 bits each. With strings this long, it took the GA a very long time to develop an effective strategy. Even after 500 generations, the results were still not suitable. After the program was modified to make simple decisions by making winning moves and preventing the opponent from winning, the string length was reduced to 104 bits. This reduced the number of generations significantly. Just after 100 generations, the GA was able to play the game at "expert" level. Therefore, by incorporating traditional AI search methods (simple searches that do not take a lot of computation power) into the GA, the program was much more effective. This lends credence to the old axiom: "If you have domain knowledge, then use it."

One of the most difficult problems in AI is the credit assignment and evaluation definition. To decide how much credit to assign to different situations is not a trivial task. Most of the time, to one extent or another, it is a trial and error process. By using GAs, the amount of work required for these problems seems to be substantially reduced. As seen from this effort, the GA does not replace the traditional AI search methods, rather, they complement each other.

Future Works

An interesting problem might be to take Samuel's checker playing program and incorporate the GA into the parameter searching routine. Samuel did a random search for his program. Even though it worked rather well, it took a lot of time for the search. By incorporating a GA, this search could be done more efficiently.

Also, the experience gained from this project could be used to attack a more complicated game. There are games that traditional AI methods do not play very well, such as bridge, poker, or go. These games represent difficult problems because of the enormous search space involved or because of the inexact information associated with the games. To come up with the right strings to represent and simplify those game strategies is the key to solving these problems.

REFERENCES

[1] Bagley, J. D. (1967). *The behavior of adaptive systems which employ genetic and correlation algorithms.* Doctoral dissertation, Ann Arbor: The University of Michigan.

[2] Ginsberg, M. (1993). *Essentials of artificial intelligence.* New York: McGraw Hill. pp. 49-99.

[3] Goldberg, D. E. (1989). *Genetic algorithms in search, optimization, and machine learning.* Reading, MA: Addison-Wesley.

[4] Haykin, S. (1994). *Neural networks – A comprehensive foundation.* pp. 397-434. New York: McGraw Hill.

[5] Michie, D. (1962). *Trial and error.* New York: McGraw Hill.

[6] Samuel, A. L. (1959). *Some studies in machine learning using the game of checkers.* New York: McGraw Hill.

CHAPTER 18

SIMULATION OF AN ARTIFICIAL ECO-SYSTEM USING GENETIC ALGORITHMS

Steve Wand
sewand@amp.com

ABSTRACT

Each species that exists on this planet is the product of millennia of natural se-lection. Competition for finite resources has produced varied species, many of which exhibit specialized behavior that allows them to survive. Genetic algo-rithms acting upon a randomly chosen population, and competing for finite re-sources should produce a near-maximal biomass, with several distinct species exploiting different levels of the biosystem. Utilizing reproduction, crossover, mutation, and niching operators, the coding scheme could preserve diversity in predator/prey populations and mass, while maximizing biomass and sensory performance within the population, particularly in a static environment. This chapter describes an investigation in which a genetic algorithm is used to simu-late an artificial environment in which various species compete with one an-other.

INTRODUCTION

Darwin's theory of evolution concludes that natural selection is the key factor in the origin of species. Within species, an individual that reproduces passes on its genetic characteristics. Individuals that possess 'favorable' traits are more likely to survive, hence future generations increasingly exhibit 'favorable' traits. Given time, a population's characteristics can diverge significantly from their original makeup. Examining the machinery of natural selection can lead to a keener appreciation of complex interactions that shape life. Because genetic algorithms are based upon the mechanisms of reproduction, they provide a clear analogy to how real populations can evolve over successive generations.

Field observations yield glimpses of natural selection's capacity to produce populations that fully exploit their environment. In the real world, this process takes millennia. A simulated ecosystem, with a diverse initial population, offers a means to view the effects of evolution over hundreds or thousands of genera-tions. The recombination of individuals via a genetic algorithm provide an ele-gant means of rewarding variations that maximize their environment.

This chapter will examine a simulated ecosystem of herbivores and carnivores. Each individual will have several characteristics that shall determine the relative success or failure of each organism within the environment. The GA operators of reproduction, crossover, mutation, and niching will operate on a multi-parameter coding. Organisms that can successfully adapt to their environment will be favored within the reproductive pool. 'Winning' populations will have the greatest increase in mass (numbers * size), with the population existing near the ideal carrying capacity of the environment.

LITERATURE REVIEW

There have been numerous works published on simulated organisms and eco-systems. Mechanisms of cell chemistry were examined by Rosenberg [1], which simulated enzyme reactions using genetic algorithm-like operators. The Avida simulated ecosystem shows support for the punctuated equilibrium view of evolution, as opposed to a more Darwinian gradual model of evolution [2].

An artificial life program called Tierra is used to model both small and large scale ecosystems. Tierra utilizes genetic algorithms to simulate evolutionary change[3]. The Tierra system creates a diverse population of organisms, but does not optimize resources by the population as a whole. To examine this problem, it is useful to look at models of population interaction. Two primary engines of ecological change are predation and competition. Ten components of functional response to prey and predation [4] are:

1)   The role of successful search
2)   Time of exposure
3)   Handling time (time taken to eat)
4)   Hunger
5)   Learning by predator
6)   Inhibitions of prey
7)   Exploitation
8)   Interference between predators
9)   Social facilitation
10)   Avoidance learning by prey

Each factor has sub-factors. For instance, successful search involves sensory facility, reaction distance, speed of predator, speed of prey, and capture success. Relationships are drawn between density of predators, density of resources, probability of attack, time spent in attack, expected gain, and number of attacks to derive a success ratio of predation. Smith [5] draws the conclusion that species that spend most of their time searching for food that takes little effort to capture, will be generalists (e.g., hyenas), while species that have abundant prey that takes much effort to capture will be specialists (e.g., cheetahs). Specialization leads to speciation. Sub-populations which converge at multiple 'peaks' along the spectrum of the initial population will eventually stop sharing genetic information with other sub-populations.

PROBLEM STATEMENT

The simulated ecosystem will have three components -- an environment, a randomly chosen initial population of herbivores and carnivores with varied characteristics, and a set of rules governing the success or failure of organisms within the ecosystem.

The environment has two component values, carrying capacity and flora color. Carrying capacity refers to the kilograms of vegetative matter available for consumption by herbivores, and for purposes of the simulation is the product of random fluctuations of rainfall and temperature. Flora color refers to the predominant shade of the vegetation, and is expressed in terms of red, green, and blue primaries that allow for colors from black to white. Carrying capacity or flora color may be varied during the course of the simulation.

The population will consist of randomly chosen individuals with the following coding scheme:

Food Source -- 0(Herbivore), 1(Carnivore)	1 byte	
Ideal Body Weight of adult -- 5 to 500 kilograms		5 bytes
Color -- Red, Green and Blue values from 0 to 256		6 bytes
Number of Legs -- 2 or 4 legs		1 byte
Vision -- Ranging from 0(poor) to 7 (great)		3 bytes
Hearing -- Ranging from 0(poor) to 7(great)		3 bytes
Brain Size -- Ranging from 0(minimal) to 31(human-like)	5 bytes	
	Total:	24 bytes

Individuals are allowed to mate freely among all members of their respective herbivore or carnivore population, subject to the constraints of the fitness evaluation algorithm.

The fitness algorithm measures the competitiveness of an individual, measured against his peers. In predator/prey systems, prey animals in the absence of predators will show a proportional growth rate. Predators introduced into a prey-rich environment will show a high growth rate, and will slow the prey growth rate. An overabundance of predators will lead to declining numbers of prey, which will, in turn, reduce the number of predators. Lotka-Volterra's equations illustrate the basic growth relationships between numbers of herbivores (H) and carnivores (C) [5]:

$$dH/dt = aH - bHC$$

$$dC/dT = -cC + dHC$$

where HC is the success rate of predation, and a,b,c,d are proportionality constants. HC, as defined by Kitching, is a function of several parameters such as detection success, learning by prey and predator, and hunger[4,5]. Therefore, our fitness algorithm must take into account the factors that lead to the success of predation. Four different functions are used to determine individual success.

(1) Herbivore feeding requirements – Herbivores require an amount of food proportional to their mass, $m^\wedge.75$. Food-rich environments favor larger animals, while food-poor environments will favor smaller animals. Accordingly, the success rate of foraging is:

$$f = (K / Eh)*(b * m^{0.75})$$

K	= the carrying capacity of the environment
Eh	= energy requirement of all herbivores
b	= proportional constant
m	= the mass of the individual

This function determines the ability of the individual herbivore to successfully forage for food. Animals that are well fed are less likely to be caught by predators.

(2) Carnivore feeding requirements - Carnivores also require food proportional to their mass, $m^{0.75}$. Because they can convert 10% of herbivore mass into energy, their success rate for foraging is:

$$f = ((Mh/10) / Ec) * (d * m^{0.75})$$

Mh	= Mass of herbivores
Ec	= Energy requirements of all carnivores

d is a proportional constant

Similar to herbivores, this function also shows the success that a well fed predator is likely to have, vis-à-vis, his starving brethren.

(3) Detection success ratio - The ability of an animal to make or escape detection depends upon their sight (s), hearing (h), and camouflage (c) compared to their opponents.

$$\text{Herbivore} = x(s / S) + y(h / H) + zc$$
$$\text{Carnivore} = i(s / S) + j(h / H) + kc$$

S, H are mean population values for sight and hearing.

i,j,k,x,y,z are proportional constants

These factors play into the role of successful search and time of exposure. Again, both equations show how an individual must do better not only in absolute terms, but in terms of the competition.

(4) Intelligence success ratio - The intelligence of an individual relative to their opponents. Highly intelligent bipedal organisms are given additional credit for tool-use capability.

$$i = (b \, / \, B)(gc)$$

b is brain size of individual

B is average brain size of population

g is a proportionality constant

c is 1 for two legged animals, 0 for four legged animals

Intelligence is a grab-bag of all of the individual's various abilities to out-wit the opponent.

The sum of the criteria produces the fitness value for the organism. The fitness value reflects the innate ability of an individual to survive the effects of environment and predation. For herbivores, this means the ability to avoid predators while successfully competing for limited forage resources. Carnivore fitness is measured by the ability to catch the prey and fend off fellow carnivores. Once determined, a stochastic remainder selection utilizes the fitness values to reproduce and crossover the fittest individuals. A mutation operator randomly modifies alleles during crossover.

The population size will be determined each generation. First, the carrying capacity of the environment is determined for each round - simulating the random effects of sunshine and rainfall. This value, taken in conjunction with the total mass of herbivores from the prior round, will determine the current round's population size. Reproduction and crossover will fill all of the available slots in the population according to the dictates of the fitness function for herbivores and carnivores.

Because the amount of food available to sustain the population is variable, particularly for carnivores, wild swings in food supply could wipe out a particular food source. A phenotypic speciation operator smoothes fitness functions based upon food source and mass. Animals with different food sources are considered dissimilar. Mass is apportioned with a linear smoothing. This accords well with common sense, since mating individuals that range in size from 5 to 500 kilograms which might be inclined to eat the mate are unlikely at best. The expectation is that species should form based upon food source and mass. Within each species, there should be improvement in the quality of its individuals, with optimizations of prey and predator near the carrying capacity of the environment.

RESULTS

The following parameters were used in all tests:

Initial Population size	25
Generations	100
Prob. of Crossover	.6
Prob. of Mutation	.001

Two series of tests were conducted. In the first test, the carrying capacity remained constant, which led to a static population size. In the second test, the carrying capacity could vary as much as 66% from one generation to another, with the population size varying as well.

For the first test, the carrying capacity remained constant at 1500 kg, and the population size was 25. For all individuals, the allele values were randomly selected. Figure 18.1 tracks the absolute variance between ideal total mass and actual total mass of herbivore and carnivores over 100 generations. Over succeeding generations, the population converged to approximately 1% variance, while allowing for crossbreeding of herbivores and carnivores.

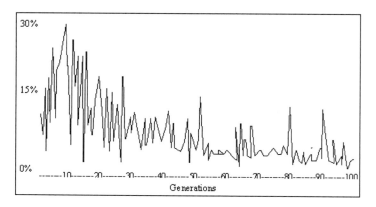

Figure 18.1. Actual variance from ideal total mass for static environment.

The average fitness value of all generations can be seen in Figure 18.2. A 400% improvement in total fitness occurred over 100 generations, due to maximizations of values for sight, hearing, color and brain size.

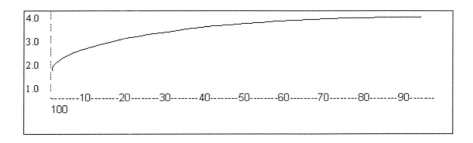

Figure 18.2. Average fitness value for static environment.

The speciation algorithm helped to maintain a relatively static number of carnivores and herbivores, with the quantity of either food type fluctuating between 8 and 17 for most of the simulation. At 100 generations, the range of individual's mass still ran from 160 to 420 kg. The average mass had gone from 260 kg initially, to 320 kg at the end, reflecting the higher mass requirements of the herbivore population. Carnivores, which depended on herbivores for their food, tended to remain at a lower weight.

The average value for vision went from 3.7 to 7. The average value for hearing went from 2.7 to 6.7. Brain size went from 13.4 to 29.2. Color went from an even distribution to varying shades of green. The number of legs per individual went from an even distribution to 90% 2-legged, reflecting greater tool use.

The second series of tests examined various carrying capacities and population sizes. The carrying capacity was allowed to range over 1000-1500 kg for each generation. The population size was incremented or decremented in proportion to the carrying capacity. Again, allele values were selected randomly. Figure 18.3 tracks the absolute variance between ideal total mass and actual total mass of herbivore and carnivores over 100 generations. Over succeeding generations, the variation did evidence some smoothing, but remained overall at approximately the 10% level.

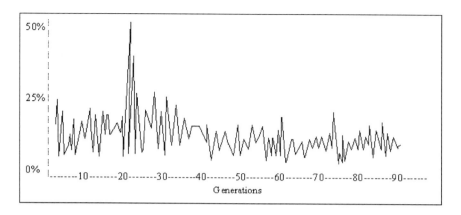

Figure 18.3. Actual variance from ideal total mass for varying environment.

The average fitness value of all generations can be seen in Figure 18.4. A 350% improvement in total fitness occurred over 100 generations, due to near maximizations of values for sight, hearing, color and brain size.

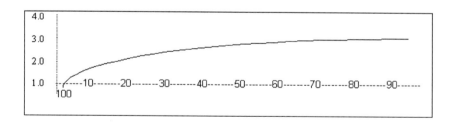

Figure 18.4. Average fitness value for static environment.

The speciation algorithm helped to maintain the relative proportion of carnivores and herbivores, although variances were higher than in the static algorithm. Because the food requirements had a much greater range, the mass of individuals also evidenced a greater range, from 20 to 420 kg. The average mass, which started at 260 kg, finished at 275 kg. The distribution contained substantial accumulation at either extreme.

The values for sight, hearing, brain size and color mimicked the results of the static run, showing good optimization performance.

CONCLUSIONS

The genetic algorithm was able to optimize the static environment's biomass much better than in the dynamic environment. Because herbivores had a constant amount of food to draw from, the optimization routine narrowed the mass range of the population considerably. This would indicate that less speciation took place in the static environment.

The relatively poor optimization of biomass in the dynamic environment could be expected, because of the additional uncertainty introduced into the herbivore food chain. The algorithm did show robustness by quickly recovering from swings in the food supply, due to the wide range of body mass preserved by the speciation algorithm. This evidence would seem to accord well with observations [5] that changing environmental conditions aggravate the swings of the predator-prey cycle.

Both environments performed well in optimizing attributes unconcerned with food requirements. The values for vision, hearing, and brain size all showed advancement from average initial values to near optimal values.

Genetic algorithms can be a useful method for determining optimal biomasses within a static, and to a lesser extent dynamic, environment. The use of tools such as speciation more closely mimic natural processes, and preserve the diversity necessary for successful response to dynamic environmental changes. At the same time, a GA is able to optimize attributes that relate to an individual's fitness.

REFERENCES

[1] Rosenberg, R. (1967). Simulation of genetic populations with biochemical properties. New York: McGraw Hill.

[2] Adami, C. T., Brown, M. & Haggerty, J. (1995). Abundance distributions in artificial life and stochastic models: "Age and Area" revisited, *Proceedings of ECAL 95*.

[3] Ray, T. (1995). Artificial Life. *ATR Human Information Processing Research Laboratories*.

[4] Kitching, R.L. (1983). *Systems ecology - An introduction to ecological modeling*. University of Queensland Press.

[5] Smith, M. (1974). *Models in ecology*. Cambridge University Press, Cambridge.

INDEX

A

ADA, 64
Alabama Cyrogenic Engineering, 75
aligned ellipses, 119
aligned single-point crossover, 167, 180
AMP Incorporated, 334
approximate function evaluation, 189
argument of perigee, 136
artificial intelligence, 316, 334
AT&T, 87

B

Bama neural network, 212
Bayesian information, 200
beam search, 159
bi-elliptic transfer, 135
biomass, 336
Boeing, 30, 64
BORN, 212
boundary data, 65

C

chaos, 256
chaotic system, 256
checkers, 316
circular orbits, 132, 133, 135, 145
comet-strike, 40, 41, 48
COMOGA, 91
computer vision, 14
controller, 34
credit assignment, 301
cyclometric, 67

D

database, 170
data mining, 155

delta coding, 37

E

eco-system, 334
elliptical orbit, 131
evolutionary programming, 2, 5
evolutionary strategies, 2, 5

F

four disk model, 41
functional mapping, 194

G

game playing, 316
Gauss, 194, 240
generation gap, 37
genetic programming, 4, 267, 282, 287
geocentric equatorial coordinate system, 152, 154
graph theory, 49, 65
Gray coding 37, 40, 42, 43

H

H_2 controller, 34
H_∞ controller, 34
hill-climbing, 40, 42, 43, 159
history, 5
Hohmann transfer, 110, 124, 132, 133, 135, 144, 154
hybrid architecture, 193, 240, 247
hydrocyclone, 219, 282, 284

I

image transformation, 14, 34
integration, 240
IPP, 65